HI-TECH EQUIPMENT RELIABILITY

HI-TECH EQUIPMENT RELIABILITY

A Practical Guide for Engineers
and Engineering Managers

Dr. Vallabh H. Dhudshia

Lanchester Press Inc.
Sunnyvale
USA

© 1995 Dr. Vallabh H. Dhudshia.
All rights reserved, including the right of reproduction in whole
or in part by any means. Printed in the United States of America

ISBN 1-57321-003-X
Library of Congress No. 95-80604

Editor: John Schuler
Electronic Imaging by Costas Schuler
Printed by Patson's Press, Sunnyvale, California, USA
Binding by Stauffer Edition Binding Co., Monterey, California

Cover photograph by Novellus Systems, Inc.

Lanchester Press Inc., P.O. Box 60621, Sunnyvale, CA 94086

To my loving family
wife Manju,
daughter Neha,
and son Neel

Contents

Figures xiii

Tables xv

Forward xvii

Preface xix

Abbreviations xxi

Chapter 1
Brief History of Reliability Discipline 1

Chapter 2
Basics of Reliability Discipline 5
2.1   Objectives 5
2.2   Formal Definition of Reliability 5
2.3   Reliability Function 7
2.4   Failure Rate Function 12
2.5   Reliability Function Related to Other Probability Functions 13
2.6   Population and Samples 16
2.7   Definition and Categories of Failures 17
        2.7.1 Catastrophic, Degradation, and Intermittent Failures 18
        2.7.2 Early, Chance, and Wear-out Failures 18
        2.7.3 Critical and Non-Critical Failures 19
        2.7.4 Independent and Dependent Failures 19
        2.7.5 Relevant and Nonrelevant Failures 19
2.8   Component and System 20

Chapter 3
Reliability Metrics 23
3.1   Introduction and Objectives 23
3.2   Two Main Categories of Reliability Terms 23
3.3   Categories of Reliability Metrics 24
      3.3.1 Metrics Based on Probabilities 25
      3.3.2 Metrics Based on Mean Life 25
      3.3.3 Metrics Normalized by Life Units 26
      3.3.4 Metrics Expressed in Percentages 27
3.4   Applications of Reliability Metrics 27
      3.4.1 Desired Values 27
      3.4.2 Analytical/Theoretical Values 28
      3.4.3 Observed Values 29
3.5   Precise Use of the Reliability Metrics 32
3.6   Standardization of the Reliability Metrics 33

Chapter 4
Reliability of Systems 37
4.1   Series System 37
4.2   Parallel System 38
4.3   Standby System 40
4.4   Repairable System 41

Chapter 5
Reliability Compared With Other Disciplines 43
5.1   Quality and Reliability 43
5.2   Quality Control, Quality, and Reliability 43
5.3   Safety and Reliability 45
5.4   Maintainability and Reliability 45
5.5   Availability and Reliability 46
5.6   Which Is More Important? 47

Chapter 6

Reliability in the High-Level Equipment Performance Metrics 49

6.1   Overall Equipment Effectiveness 49
        6.1.1 Availability 50
        6.1.2 Performance Efficiency 50
        6.1.3 Quality Rate 51
        6.1.4 An Example of OEE Calculations 51
6.2   Life Cycle Cost 51
6.3   Cost of Ownership 54
6.4   Hierarchy of Equipment Performance Metrics 57

Chapter 7

Equipment and Equipment Program Life Cycle Phases 59

7.1   Equipment Life Cycle Phases 60
7.2   Equipment Program Life Cycle Phases 60
        7.2.1 Concept and Feasibility Phase 61
        7.2.2 Design Phase 62
        7.2.3 Prototype Phase 63
        7.2.4 Pilot Production Phase 63
        7.2.5 Production Phase 64
        7.2.6 Phase Out Phase 64
7.3   Applying Reliability Metrics During Equipment Life Cycle 64

Chapter 8

The Reliability Improvement Process and Its Applications 67

8.1   The Reliability Improvement Process 67
        8.1.1 Establish Reliability Goals and Requirements 68
        8.1.2 Reliability Engineering and Improvement Activities 69
        8.1.3 Conduct an Evaluation 69
        8.1.4 Determine Whether Goals and Requirements Are Being Met 69
        8.1.5 Identify Problems and Root Causes 69

8.2	Applying the Reliability Improvement Process	70
8.3	Activities Associated with the Reliability Improvement Process	70
8.4	Reasons to Use a Reliability Improvement Process	71
8.5	Reliability Plans	71
	8.5.1 General Company Level Reliability Plan	*72*
	8.5.2 Product Level Reliability Plans	*72*

Chapter 9
Four Steps to Better Equipment Reliability ... 75

9.1	Know Goals and Requirements	75
	9.1.1 Goal Allocation	*76*
9.2	Design in Reliability	77
	9.2.1 Use Proper Parts Correctly	*79*
	9.2.2 Use Proper Design Techniques	*81*
	9.2.3 Minimize Effect of External Factors	*81*
	9.2.4 Avoid Failures Through Scheduled Maintenance	*82*
	9.2.5 Design Review	*83*
	9.2.6 Reliability Assessment of the Design	*86*
9.3	Build-in Reliability	87
9.4	Manage Reliability Growth	87

Chapter 10
Three Reliability Growth Mechanisms ... 91

10.1	Reliability Growth Mechanism During the Early Life of Equipment	92
10.2	Reliability Growth Mechanism Throughout the Equipment Program Life Cycle Phases	93
10.3	Reliability Growth Mechanism From One Generation to the Next Generation	96

Chapter 11
Reliability Testing 97
- 11.1 Types of Reliability Tests 98
  - *11.1.1 Burn-in Test* 98
  - *11.1.2 Environmental Stress Screening Test* 98
  - *11.1.3 Reliability Development/Growth Test* 99
  - *11.1.4 Reliability Qualification Test* 99
  - *11.1.5 Product Reliability Acceptance Test* 99
  - *11.1.6 Accelerated Test* 99
- 11.2 Generic Steps for Reliability Tests 100
  - *11.2.1 Test Plan Development* 100
  - *11.2.2 Test Conducting* 101
  - *11.2.3 Test Data Analysis and Reporting* 101
- 11.3 Reliability Tests Throughout the Equipment Program Life Cycle Phases 101
- 11.4 Test Length 102
- 11.5 Test Data Analysis 103

Chapter 12
How to Buy Reliable Equipment and Parts 105
- 12.1 Select Proper Supplier 106
- 12.2 Communicate Reliability Requirements 107
- 12.3 Buy with Guarantee and Maintenance Contract 108
- 12.4 Insist Upon a Reliable Product 108
- 12.5 Form Partnership with Suppliers 108
- 12.6 Provide Feedback 109

Chapter 13
Reliability Organization 111
- 13.1 Make-up of a Typical Reliability Organization 111
  - *13.1.1 Executive Champion* 112
  - *13.1.2 Technical Champion/Reliability Manager* 112

    13.1.3 Reliability Engineer 113
    13.1.4 Reliability Technician 113
13.2 Organization Structure 113
13.3 Recommended Practices for Reliability Engineers 114

Chapter 14
How Good Is Your Reliability Improvement Program? 117
14.1 Evaluation Methodology 117
14.2 Evaluation Process Steps. 120

Appendix 127
Index 133

Figures

1.1 History of Reliability Discipline 4
2.1 Light Bulbs Life Test 8
2.2 Graphs of Reliability Function R(t) and 1-R(t) 12
2.3 Failure Rate Graph 13
2.4 Exponential PDF for MTBF = 100 Hours 15
2.5 Exponential CDF and Reliability Function for MTBF = 100 Hours 15
2.6 Well Known PDF's and Reliability Functions 16
2.7 Population and Sample Relationship 17
2.8 Early, Chance, and Wear-out Failures 18
3.1 Reliability Metrics Based on Mean Life 25
3.2 Reliability Metrics Normalized by Life Units 26
3.3 Applications of Reliability Metrics Originated From Desired Values 29
3.4 Applications of Reliability Metrics Originated From Theoretical/
    Analytical Values 30
3.5 Applications of Reliability Metrics Originated From Observed Values 31
3.6 Reliability Metrics and Their Applications 32
3.7 Equipment States of SEMI E10–92 34
3.8 Key Formulas of SEMI E10–92 35
4.1 Series System 38
4.2 Parallel System 39
4.3 Standby System 40
4.4 Repairable System as a Superimposed Renewal Process 41
5.1 Relationship Among Quality, Reliability, and Safety 44
5.2 Availability vs. MTBF 46
6.1 A Typical MTBF vs. OEE Graph 52

6.2 A Typical Reliability Level vs. Life Cycle Cost 53
6.3 Hierarchy of High-Level Equipment Performance Metrics 57
7.1 Equipment Program Life Cycle Phases in Sequential and
     Concurrent Formats 61
7.2 Proper Uses of Reliability Metrics Throughout Equipment
     Life Cycle Phases 65
8.1 The Reliability Improvement Process 68
9.1 Process of Designing-in Reliability 78
9.2 FRACAS Process Flow 88
9.3 A Typical Failure Report Form 89
10.1 Hierarchy of the Reliability Growth Mechanisms 92
10.2 Reliability Growth Mechanisms 94
12.1 Buying Process with the Reliability Improvement Activities 106
13.1 Typical Organization with a Reliability Engineering Group
     (Organization) 114
14.1 Reliability Program Evaluation Guide 126

Tables

2.1 Commonly Used Measures of Life and Their Units 7
2.2 Light Bulb Life—Test Data. 9
2.3 Sorted Light Bulb Failure Times 10
2.4 Reliability Function For Light Bulbs 11
3.1 Two Main Categories of Reliability Terms. 24
6.1 A Typical Life Cycle Cost Calculation 54
6.2 A Typical Cost of Ownership Calculation 56
8.1 List of Reliability Improvement Activities Throughout
    Equipment Life Cycle Phases. 73
9.1 Derating Factors for Electronic Parts. 80
9.2 Design Review Team Members and Their Responsibilities 84
9.3 System Level Checklist for Design Review Team 85
9.4 Part Level Checklist for Design Review Team. 86
10.1 Recommended Values of Reliability Growth Constant α 95
11.1 Reliability Tests Throughout the Equipment Life Cycle Phases. 102
11.2 Minimum Test Length Multiplier ϖ 103
11.3 Multiplier K for the Confidence Limit Calculations. 104
14.1 Reliability Program Evaluation Checklist—Company Culture
    Related Activities 119
14.2 Reliability Program Evaluation Checklist—Reliability Group
    Related 120
14.3 Reliability Program Evaluation Checklist—Reliability Goals/
    Objective Related Activities. 121
14.4 Reliability Program Evaluation Checklist—Design Assurance
    Activities 122

14.5 Reliability Program Checklist—Reliability Testing Tasks 123
14.6 Reliability Program Evaluation Checklist—Manufacturing
      Quality Assurance Tasks..124
14.7 Reliability Program Evaluation Checklist—Reliability Growth
      Management Tasks 125
A1. Concept and Feasibility Phase: Activities by Reliability
      Improvement Process Step 127
A2. Design Phase: Activities by Reliability Improvement
      Process Step 128
A3. Prototype Phase: Activities by Reliability Improvement
      Process Step 129
A4. Pilot Production Phase: Activities by Process Improvement Step.........130
A5. Production and Operation Phase: Activities by Reliability
      Improvement Step...131
A6. Phase-Out Phase: Activities by Reliability Improvement Step 132

Forward

Reliability and Product Assurance Engineers need to design, manufacture, test and deliver to the customer products with optimum reliability (minimum life-cycle cost), which are easy to maintain, safe to operate, of highest quality, and sold at competitive prices. This book tells the Reliability Engineer and the Reliability Engineering and Product Assurance Manager and Director, in a concise manner, all the necessary aspects of this important subject area.

Included are: basic definitions, reliability metrics, comparison with other disciplines, equipment program life cycle phases, reliability improvement activities throughout the life-cycle phases, four basic steps to better equipment reliability, reliability growth mechanisms, reliability testing and reliability organization.

This book also outlines a systematic approach to buying reliable parts, and a step-by-step procedure to evaluate the effectiveness of an organization-wide reliability improvement program. It should be in the library or bookcase of every Reliability Engineering practitioner.

Professor Dimitri Kececioglu
The University of Arizona

Preface

Reliability has been widely used to measure equipment performance in military and commercial industry since the early 1940's. Movements to track high-level matrices, such as Overall Equipment Effectiveness or Cost of Ownership, are more recent developments. Since all such matrices rely heavily on a reliability metric, they do not dilute importance of the reliability discipline. On the contrary, they enhance it.

Today's highly competitive, global market environment demands an optimum level of reliability in present and future products/equipment. Customers expect and, in some cases, competitors force a high reliability level from manufacturing organizations. At the same time, the complexity of most equipment is continuously increasing. These influences are driving ever-higher reliability modules and components just to maintain the same reliability level.

To make a reliability improvement program effective and achieve a world-class reliability level, everyone in the organization—not just the reliability engineer—must play his or her part. Usually, however, not everyone is equipped with adequate knowledge of the discipline to play the part effectively. Reliability engineers understand the implications of this trend. An abundance of available textbooks, military handbooks, standards, and guidebooks use high-level mathematics and statistical theory to help define and clarify reliability discipline for reliability engineers. However, this reliability discipline needs to be understood and applied by everyone in an organization, not just reliability engineers.

Hi-Technology Equipment Reliability targets design, manufacturing, and service engineers; management; purchasing staff; and all others within an organization who can actively contribute product/equipment reliability. Great emphasis is placed on gaining a basic working knowledge of reliability discipline and understanding the basic phases of a reliability improvement program throughout a product's developmental life cycle. Examples from actual industry experience enhance the usefulness of the topic. This book focuses on equipment reliability for the semiconductor manufacturing industry; however, the material presented can be applied to any product or equipment line. Some knowledge of mathematics would be helpful but is not absolutely necessary to understand the subject matter.

This book will not make a reliability engineer out of you. It will, however, help you become a knowledgeable partner with reliability engineers and others within your organization to help your company reach its reliability goals.

The content of this book is based on my life experiences during my association with SEMATECH as an assignee from Texas Instruments, Inc.; first helping semiconductor manufacturing equipment suppliers implement reliability improvement programs and, second, developing and teaching a class entitled "Equipment Reliability Overview for Design, Manufacturing, and Field Engineers, Purchasing Personnel, and Project and Program Managers" to the same suppliers. The knowledge I gained during the above interactions is included in this book.

I would like to thank Texas Instruments Inc., for letting me serve SEMATECH and for permitting me to share my work in book form. In addition, I would like to thank SEMATECH for giving me an opportunity to work on the subject matter and for permitting me to write this book. This acknowledgment is not complete without thanking staff members of the External Total Quality and Reliability group at SEMATECH, who have helped me to continuously improve the subject matter.

I am also deeply indebted to Sheila Endres and Ken Griger for their editorial assistance in preparing the manuscript.

Finally, I offer my sincere gratitude to Lanchester Press for encouragement to publish this book and for all their kind publishing assistance.

Dr. Vallabh H. Dhudshia

Abbreviations

AGREE	Advisory Group on Reliability of Electronic Equipment
AITPM	American Institute for Total Productive Maintenance
ARINC	American Radio Inc.
ASQC	American Society for Quality Control
CDF	Cumulative Density Function
CoO	Cost of Ownership
CT	Cycle Time
ESS	Environmental Stress Screening
FMEA	Failure Mode and Effects Analysis
FRACS	Failure Reporting, Analysis, and Corrective Action
FRB	Failure Review Board
FRU	Field Replaceable Units
FTA	Fault Tolerance Analysis
Hr	Hour or Hours
IEEE	Institute of Electrical and Electronic Engineers
ISO	International Organization for Standardization
LCC	Life Cycle Cost
MCBF	Mean Cycles Between Failures
MTBF	Mean Time Between Failures
MTTR	Mean Time to Repair

OEE	Overall Equipment Effectiveness
PDF	Probability Density Function
PM	Preventative Maintenance
PO	Purchase Order
PPH	Parts per Hour
PRAT	Product Reliability Acceptance Test
QC	Quality Control
RFQ	Request for Quotation
RH	Relative Humidity
RIP	Reliability Improvement Process
SEMI	Semiconductor Equipment and Materials International Inc.
SM	Scheduled Maintenance
SSQA	Standardized Supplier Quality Assessment
TPM	Total Productive Maintenance

Chapter 1
Brief History of Reliability Discipline

Early in the 1940's, equipment manufacturers recognized the need for an independent reliability discipline. At that time, emphasis was on the reliability of electronic parts for military equipment. During this same time period, the U.S. Departments of the Army and the Navy developed parts standards for a few critical electronics parts.

In the 1950's and 1960's, the discipline saw phenomenal growth on every front. Reliability reached a very high level of awareness, and its influence became widespread.

By the 1950's, the reliability discipline had expanded to include military electronic equipment and systems. Reliability engineering started to become an important and independent discipline. Leading corporations began establishing formal programs for reliability discipline, and they drew the attention of professional societies such as Institute of Electrical and Electronics Engineers (IEEE) and the American Society for Quality Control (ASQC). Organizations started holding separate symposia on reliability discipline. At the same time, Advisory Group on the Reliability of Electronic Equipment (AGREE) was established and became very active in providing direction and guidelines for the discipline, both for military equipment and for commercial products. By the late 1950's, textbooks on reliability started appearing in bookstores.

As the issue of reliability came to the forefront of management thinking, manufacturers began to become aware of the need for reliability program management throughout product life cycle phases.

By the early 1960's, the U.S. Army and Navy were teaching formal courses in reliability engineering. By the mid–1960's, many textbooks on reliability statistics and engineering were published, and U.S. universities started teaching formal courses in reliability engineering. Additional professional societies were established to cater to the growing interest. At this time, reliability discipline began including the reliability of mechanical parts.

In the 1970's and 1980's, reliability discipline spread throughout most commercial products. Reliability improvement programs became more formal, well organized, and documented. High reliability level became one of the customer's requirements. It also became an instrument for manufacturers to compete in the global market place. Focus was also directed on the cost of achieving a reliability level. This led to life cycle cost and optimum reliability level with minimum life cycle cost. During this time, equipment and systems became more complex, and their operations became more software dependent. These changes led to a need for reliable software. As a result, software reliability emerged as a part of the reliability discipline.

In the late 1980's, the focus of reliability improvement efforts shifted from screening and inspecting to proactive activities such as designed-in, built-in, and managed growth. This change moved many reliability improvement activities to the beginning of a product's life cycle, at the design and development phase. It was during this time, too, that manufacturers realized their parts suppliers and customers could play an important role in achieving reliability goals. Instead of being simply their supplier's customer and their customer's supplier, they started partnering and cooperating with suppliers and customers for mutual gain. This partnership led to an early involvement of parts suppliers and product customers at the design and development stage to implement reliability improvements.

Within their organizations, manufacturers realized that achieving a desired level of reliability is not only the responsibility of reliability engineers, but required involvement of design, manufacturing and field service engineers; marketing; purchasing; and program managers. High level corporate managers recognized the need for reliability discipline and provided the needed resources. They made the reliability discipline a part of their business systems and incorporated formal reliability improvement plans in the business plans.

The relationships between suppliers and customers became even stronger in the early 1990's. Both initiated more joint development programs, increased cooperation, and exchanges of information. Suppliers started having access to real-life experience data for the parts or equipment they supplied. Manufacturers started reducing their supplier base and became selective when choosing suppliers. Supplier assessment and certification became prerequisites before manufacturers would deal with them. Existing reliability improvement programs became an important criterion for selecting and certifying a supplier. Reliability requirements became an essential part of an equipment purchase process.

In the early 1990's, some equipment users began tracking high level metrics, such as Overall Equipment Effectiveness (OEE) and Cost of Ownership (CoO). But reliability remains a key element of such metrics; therefore, emphasis on reliability discipline is ever increasing.

Figure 1.1 shows the history of reliability discipline in a graphic format.

1990s
Partnership Forms Between Customers and Suppliers.
Supplier Certification and ISO 9000 are Used.
High-level Reliability Metrics are Tracked.

1980s
Reliability Becomes a Customer Requirement.
Emphasis Shifts to Proactive Activities.
Global Competition Forces Reliability Improvements.
Software Reliability Emerges.

1970s
Application Spreads to Customer Products.
Reliability Programs Become More Formal and Well Documented.
LCC Becomes a Part of Reliability Programs.
Mechanical Part Reliability Becomes Active.

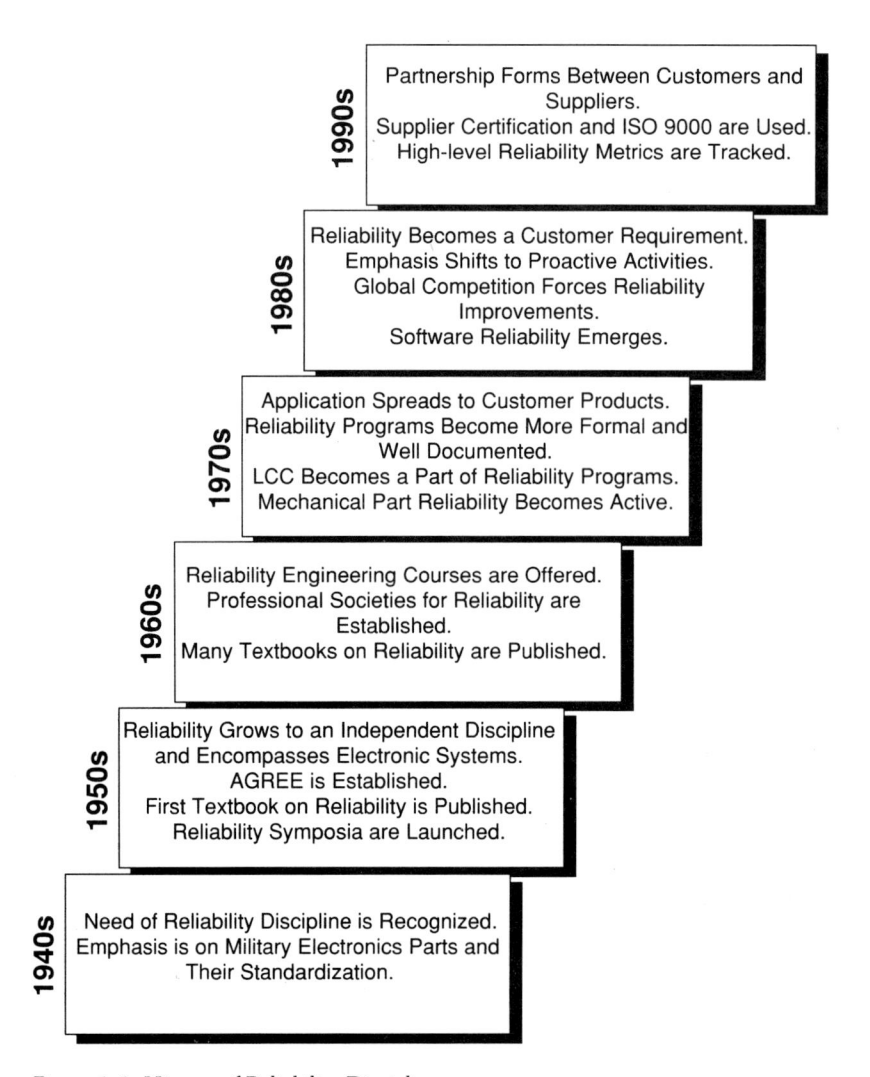

1960s
Reliability Engineering Courses are Offered.
Professional Societies for Reliability are Established.
Many Textbooks on Reliability are Published.

1950s
Reliability Grows to an Independent Discipline and Encompasses Electronic Systems.
AGREE is Established.
First Textbook on Reliability is Published.
Reliability Symposia are Launched.

1940s
Need of Reliability Discipline is Recognized.
Emphasis is on Military Electronics Parts and Their Standardization.

Figure 1.1 History of Reliability Discipline

Chapter 2
Basics of Reliability Discipline

2.1 Objectives

The objective of this chapter is to present a simple definition for reliability and for the following five basic elements of reliability discipline:

1. Reliability function
2. Failure rate function
3. Population and samples
4. Failure and its categories
5. Component and system—non-repairable and repairable

Other prime objectives of this chapter are to show relationships between reliability function, failure rate function, and the two most popular statistical functions: cumulative density function (CDF) and probability density function (PDF).

2.2 Formal Definition of Reliability

Reliability is the probability of performing intended functions for a specified time under the stated operational conditions.

Mathematically, it is written as:

$$R(t) = \Pr[T > t] \tag{2.1}$$

WHERE:

t = Specific time of interest
T = Random variable
R(t) = Reliability at time t
Pr[ ] = Probability of

EXAMPLE:

$$R(1000Hr) = Pr[T > 1000Hr] = 0.95$$

In this example, 95% of the equipment[1] units should survive past 1000 hours.

Three key points of the above formal definition require further explanation.

Intended Functions. Every equipment has its intended functions, whether they are formally documented or not. However, a given reliability level applies to a given set of the functions that the equipment was designed to accomplish. If the equipment is used for functions other than its intended design, the same reliability level may not apply to these new functions. It is the manufacturer's responsibility to see that equipment users understand the equipment's intended functions.

Specified Time. The reliability level changes as the equipment ages. It is necessary to include equipment age in establishing a reliability level. Without inclusion of such time element, any reliability level is ambiguous and can mislead a user about the specific reliability level.

Stated Operational Conditions. Factors such as operating environment, operating stress level, operating speed, operator skill level, and maintenance procedures and policies can affect the reliability of any equipment. If the value of any factor varies from assumed operational conditions, the reliability level may differ.

EXAMPLE:

The reliability of a blower in a card cage operating in its ambient environment at 60% of its rated power will be 0.85 at 2 years after installation.

---

1 In this book the words "equipment," "product," and "system" mean more or less the same thing. For uniformity, "equipment" is typically used. In some places, the word "product" or "system" is used to enhance readability. However, the words are used interchangeably.

2.3 Reliability Function

Reliability function is defined as:

$$R(t) = N_s(t)/N \tag{2.2}$$

WHERE:

N = Number of identical product units put on a life test
$N_s(t)$ = Number of units survived after time t
t = Specific time of interest

For the illustration purpose, *time* is used as a measure of life in this chapter. As shown in Table 2.1, many other measures of life are used in industry.

Table 2.1 Commonly Used Measures of Life and Their Units

Measures of Life	Units
Time	Hours, Days, Years, etc.
Number of Cycles Performed	Cycles, K-Cycles, etc.
Number of Wafer Processed	Wafers
Distance Traveled	Feet, Miles, etc.
Number of Prints Made	Prints
Number of Transactions Made	Transactions

The above equation leads to a simple methodology to develop the reliability function for any real-life situation.

1. Put N identical product units on a life test under identical operating stresses and operating environments. Figure 2.1 illustrates such a life test for 100 identical light bulbs.

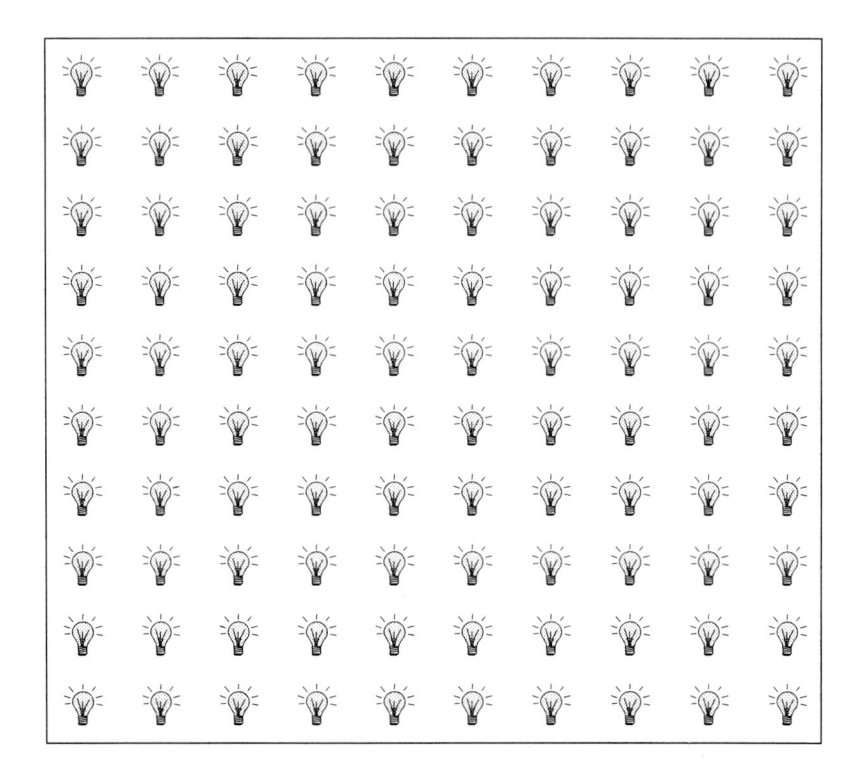

Figure 2.1 Light Bulbs Life Test

2. Start the life test and keep a record of the failure time (life length) of each unit as shown in Table 2.2.

Table 2.2 Light Bulb Life—Test Data

	Observed Time to Failure, Hour
487	256	104	757	86	637	137	54	29	30
333	86	14	542	29	140	190	48	79	65
1	553	54	2	18	84	80	79	125	301
433	39	113	426	166	78	512	591	117	497
58	255	23	179	359	146	499	153	695	100
125	45	460	33	454	33	13	137	261	374
297	357	5	316	222	551	174	3	139	33
20	60	62	244	95	359	267	748	222	162
22	69	108	90	199	12	79	208	21	72
32	29	287	13	661	198	94	128	14	241

3. Sort the failure time data collected in Step 2 from the smallest to the largest, as shown in Table 2.3

Table 2.3 Sorted Light Bulb Failure Times

\multicolumn{10}{c	}{Time to Failure, Hour}
1	20	33	65	86	125	174	255	359	512
2	21	33	69	90	128	179	256	359	542
3	22	39	72	94	137	190	261	374	551
5	23	45	78	95	137	198	267	426	553
12	29	48	79	100	139	199	287	433	591
13	29	54	79	104	140	208	297	454	637
13	29	54	79	108	146	222	301	460	661
14	30	58	80	113	153	222	316	487	695
14	32	60	84	117	162	241	333	497	748
18	33	62	86	125	166	244	357	499	757

4. Select at least ten equal-length intervals within the range and then determine the number of units that survived time t.
5. Calculate R(t) for each selected t.

6. Prepare a table of t and R(t) similar to Table 2.4, and also draw graphs of t vs. R(t) and t vs. 1 − R(t) similar to those shown in Figure 2.2.

Table 2.4 Reliability Function For Light Bulbs

Light Bulb Age t, Hour	Reliability Function R(t) at Age t
0	1.00
20	0.89
40	0.77
60	0.71
80	0.62
100	0.55
120	0.51
140	0.44
160	0.42
180	0.38
200	0.35
220	0.34
240	0.32
260	0.28
280	0.26
300	0.24
320	0.22
340	0.21
360	0.18
380	0.17
400	0.17

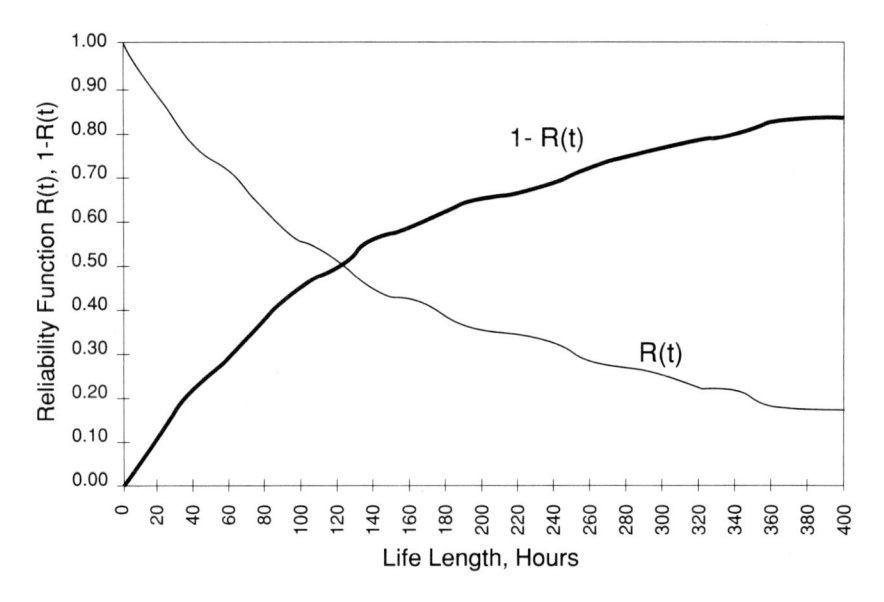

Figure 2.2 Graphs of Reliability Function R(t) and 1-R(t)

For all practical purposes, the t vs. R(t) graph is a powerful tool for non-reliability engineering personnel. This graph can be used to interpolate reliability at any time of interest without going into reliability mathematics. For example, the reliability of the light bulb is 0.65 at 70 hours.

2.4 Failure Rate Function

The second most important function in reliability is the failure rate function, $\lambda(t)$, which relates the age t of a product unit and the number of failures per unit time at that age. Mathematically, it is given as:

$\lambda(t)$ = (Number of units failed between $t - \Delta t/2$ and $t + \Delta t/2$)/
(Number of units at the beginning of the age interval x Δt)   (2.3)

WHERE:

$\lambda(t)$ = Value of the failure rate function at time t
Δt = Age interval increment

Figure 2.3 shows the $\lambda(t)$ vs. t graphs for the light bulb's life test data, which appears to be a straight line.

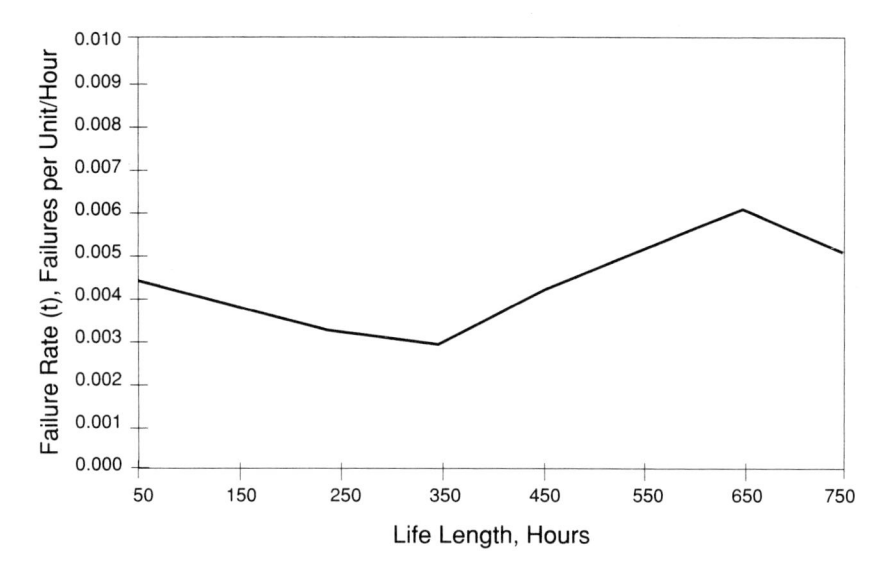

Figure 2.3 Failure Rate Graph

2.5 Reliability Function Related to Other Probability Functions

Two widely used probability functions are cumulative density function (CDF) and probability density function (PDF). See Reference 1 for a detailed statistical definition.

CDF is usually noted as F(t) and it refers to cumulative failure percentage at time t. Reliability Function refers to surviving percentage at time t. Therefore, both functions are related with the following relationship:

$$1 - R(t) = F(t) \qquad (2.4)$$

As mentioned earlier, Figure 2.2 contains a graph of F(t) for our bulbs.

PDF is usually noted as f(t) and it refers to failure probability density at time t. In layman's language, it shows percentage failed in a unit of time at time t. Both CDF and PDF are very closely related, as follows:

$$f(t) = d[F(t)]/dt \text{ or } F[t] = \int_0^t f(t)\,dt \tag{2.5}$$

WHERE:

F(t) = CDF at time t
f(t) = PDF at time t

From relationships of Equations (2.4) and (2.5), it is obvious that all three functions (reliability function, CDF, and PDF) are mathematically related. We need to know only one of them to calculate the other two.

For example, the most popular, simple, and widely used PDF in reliability discipline has exponential relationship between f(t) and t. It is known as *Exponential PDF* and is given as follows:

$$f(t) = (1/\theta)e^{-t/\theta} = \lambda e^{-\lambda t} \tag{2.6}$$

WHERE:

θ = Mean life
λ = Failure rate = 1 /(mean life)

Note that the failure rate is constant for exponential PDF. Also, we need to know only mean life to characterize the entire distribution.

Using Equation (2.5)

$$F(t) = 1 - e^{-(t/\theta)} = 1 - e^{-\lambda t} \tag{2.7}$$

And using Equation (2.4)

$$R(t) = e^{-(t/\theta)} = e^{-\lambda t} \tag{2.8}$$

Figure 2.4 shows a graph of the PDF and Figure 2.5 shows graphs of CDF and Reliability Function for exponential distribution for MTBF=100 hours. Similar graphs are available for other PDFs in any statistical textbook. Figure 2.6 shows such graphs for some of the popular PDFs.

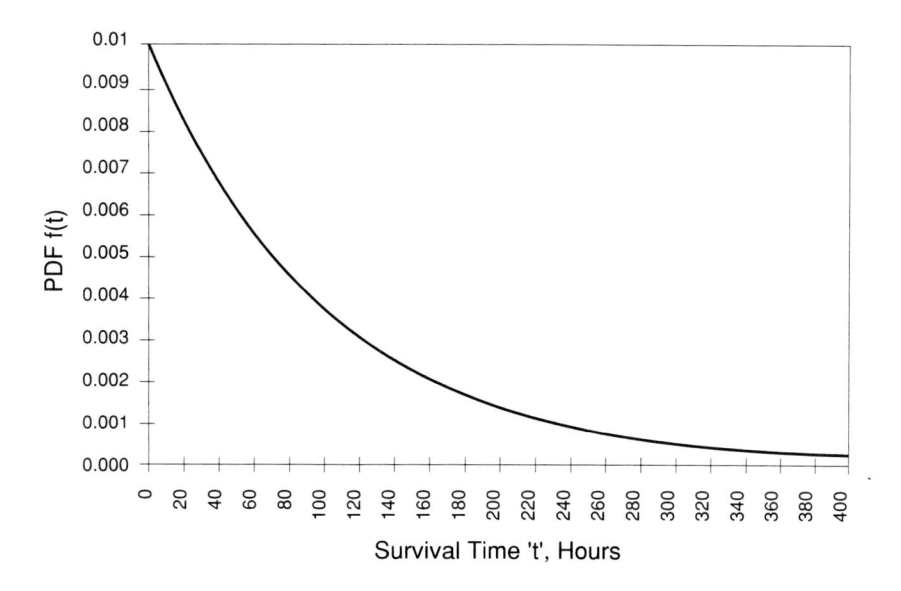

Figure 2.4 Exponential PDF for MTBF = 100 Hours

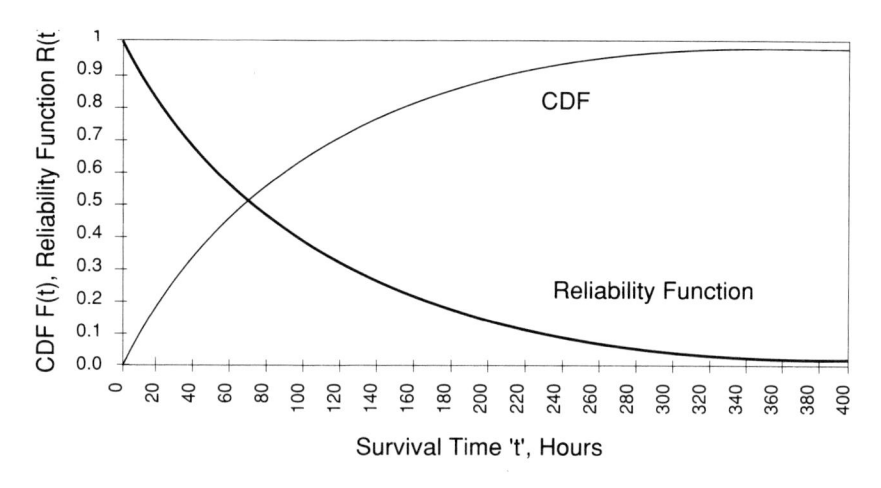

Figure 2.5 Exponential CDF and Reliability Function for MTBF = 100 Hours

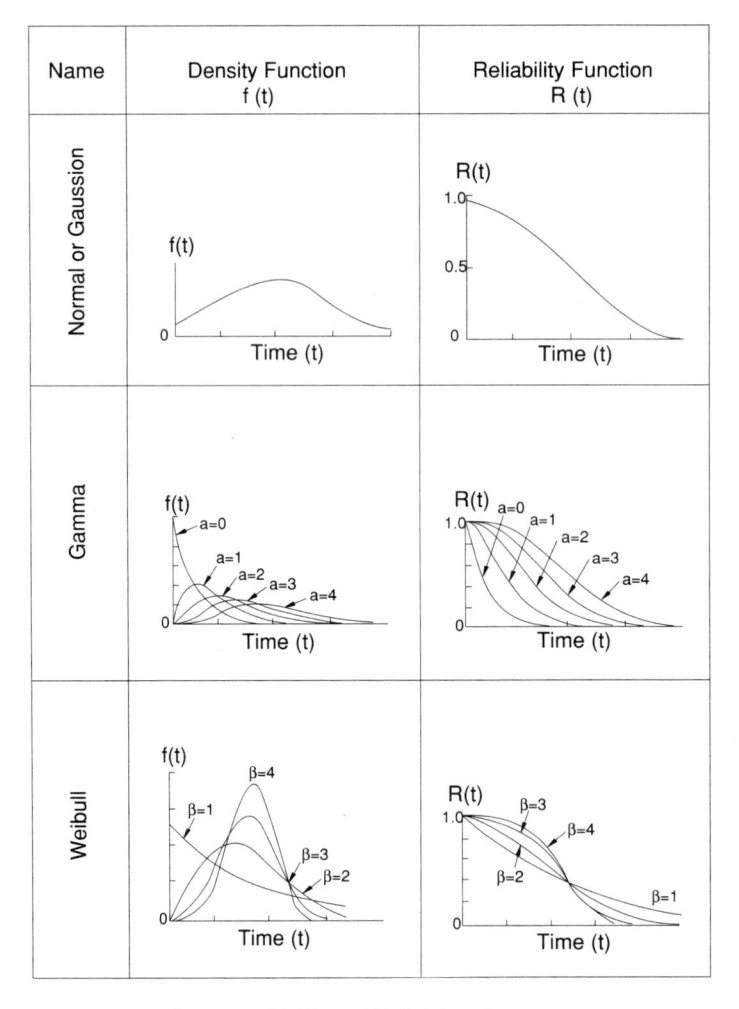

Figure 2.6 Well Known PDF's and Reliability Functions

2.6 Population and Samples

It is important that we understand the difference between a population and a sample and the statistical relationship between them. A *population* represents all the similar units performing the similar functions. For example, all the units belonging to an entire equipment line represent an equipment population. Generally, the population is large, and it is impractical to measure the value of a population metric of interest. Therefore, we randomly select a limited number of units from the population, called *samples*, to measure value of metric of interest. Since the samples are limited in number, the measurements are practical. The sample metric values are estimates of the respective population metric. In statistical terminology, the value of the population metrics are inferred from the value of the corresponding sample metrics. Statistical techniques (see Reference 2) are used to determine confidence in the inferences. The larger the sample size, the more confidence we have in the estimates for the respective population metric. Figure 2.7 depicts relationship between population and sample.

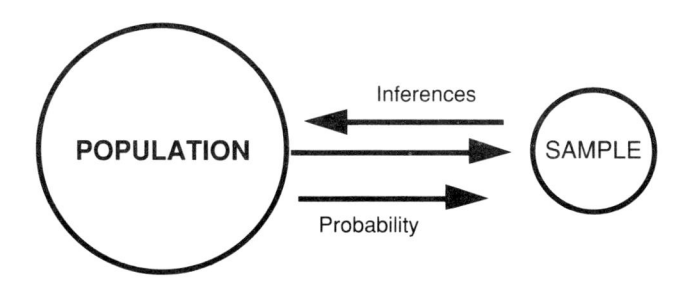

Figure 2.7 Population and Sample Relationship

2.7 Definition and Categories of Failures

Two main elements of the reliability discipline are failure and time (or other measures of life) to failure. *Failure* is defined as an event or state in which any equipment or part of the equipment does not or would not perform as intended. Some subjectivity may be required in the phrase "does not perform as intended" if the intended functions are not defined thoroughly.

Failures can be categorized in many different ways. Five of the most widely used failure categorizations are:

2.7.1 Catastrophic, Degradation, and Intermittent Failures

Catastrophic failures are sudden, unexpected, and nonreversible (e.g., broken part, short circuit, open resister, etc.)

Degradation failures occur when the output degrades slowly below the expected level and it is not revisable (e.g., bearing wear out, corrosion of surface, fatigue, etc.)

Intermittent failures are those events in which the equipment performance keeps flip-flopping below and within the expected level at unknown times and for unknown reasons (e.g., intermittent circuit board failures).

2.7.2 Early, Chance, and Wear-out Failures

Figure 2.8 shows a graph of t vs. failure rate $\lambda(t)$ for typical equipment. Since this graph looks like a cross section of a bathtub, it is known as a *bathtub* curve in the reliability world.

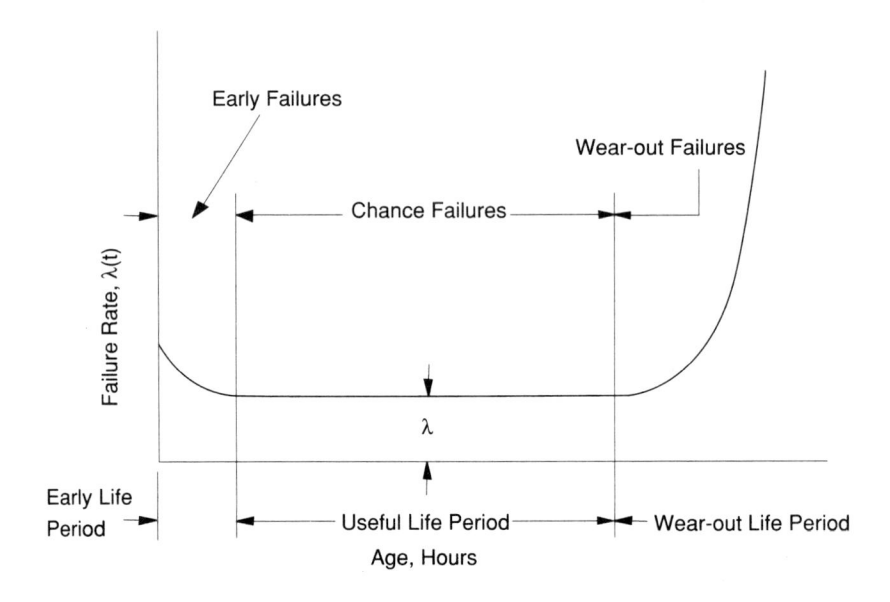

Figure 2.8 Early, Chance, and Wear-out Failures

The failures that occur during the early life period when the failure rate is decreasing are called *early failures*. These failures are caused by poor manufacturing practices, poor quality control, insufficient burning-in or screening of parts, and improper debugging after the final assembly.

Chance failures occur during the period when the failure rate is constant (middle of the graph in Figure 2.8). These failures are caused by design errors, misapplication of parts, unexplainable causes, and improper operations.

Wear-out failures occur during the period when the failure rate is increasing (see Figure 2.8) after staying constant at a lower level. These failures are caused by aging of parts, fatigue, creep, corrosion, or other deterioration caused by age.

2.7.3 Critical and Non-Critical Failures

Critical failures stop the equipment from performing the intended functions, while non-critical failures do not affect the equipment performance. For example, a label falling off is a non-critical failure. However, if a safety label falls off and the regulations require the label be affixed properly at the proper location to operate the equipment, then it is a critical failure.

2.7.4 Independent and Dependent Failures

If something fails without the influence of other failures in the equipment or outside factors, then it is an *independent failure*. If something fails because of another failure, it is considered a *dependent failure*. For example, a card cage exhaust blower failure (an independent failure) increases temperature in the card cage which causes a circuit board failure. In this situation, the circuit board is a dependent failure. Sometimes, dependent failures are discounted in reliability calculations.

2.7.5 Relevant and Nonrelevant Failures

Relevant failures are caused by failure(s) of components, modules, software, and process while performing their intended functions. *Nonrelevant failures* are caused by other factors that are not part of the equipment performing the intended functions. For example, power failure, facility problems, or an out-of-spec consumable can stop intended functions of the equipment. These failures are nonrelevant failures for the equipment operation. Nonrelevant failures are discounted in reliability calculations.

2.8 Component and System

It is also important that we understand the difference between a component and a system.

A *component* is a basic part of a system that may be an individual piece or a complete assembly of individual components. It is not subjected to disassembly and, hence, it is discarded the first time it fails. For example, a heat lamp in a heater assembly is considered a component.

A *system* is a combination of components, parts, assemblies, modules, accessories, and software connected to perform the intended functions. At least one component in the system must fail to cause the system[2] to fail. For example, a power supply failure in a tungsten deposit system may stop it from performing its intended function.

The systems belong to either one of two main categories—*non-repairable systems* and *repairable systems*. A *non-repairable system* is discarded the first time it ceases to perform its intended function(s), i.e., when it fails. A system that, after failing to perform at least one of its intended functions, can be restored to perform all of its intended functions by any method other than replacing the entire system is called a *repairable system*. A repairable system can be restored by replacing, repairing, adjusting, or cleaning the appropriate component(s).

Most large systems, such as semiconductor manufacturing equipment, are repairable systems. The distribution of failure times between two successive failures of a repairable system is discussed in Chapter 4.

There is subjectivity in the values for the above definitions. A *component* for one person may be a *system* for another. For example, for all practical purposes, a computer monitor is a component in a large process equip-

---

2 As noted earlier, the term "system" is synonymous with the term "equipment" in this book.

ment system, while it is a system for the manufacturer of the monitor. For many repairable systems, field replaceable units (FRU's) are considered components.

Three basic types of systems (series, parallel, and standby) are discussed in Chapter 4.

REFERENCES

1. Gerald J. Hahn and Samuel S. Shapiro, *Statistical Models in Engineering*, John Wiley & Sons, Inc., New York, NY, 1967.

2. William W. Hines and Douglas C. Montgomery, *Probability and Statistics in Engineering and Management Science*, John Wiley & Sons, Inc., New York, NY, 1980.

Chapter 3
Reliability Metrics

3.1 Introduction and Objectives

Experience shows that reliability metrics often uses a variety of terms that are not always precise and simple to understand. These terms mean different things to different people, and the result is confusion, irritation, and animosity. This is particularly true for nonreliability engineering personnel. The most common confusion stems from a mix-up of the reliability metrics and the application of those metrics. The purpose of this chapter, therefore, is to define and categorize the most frequently used reliability metrics and their applications and then to clarify their relationship with each other.

We will limit our discussion to exponential PDF, because failure times of almost all electronic components and that of most mechanical components follow exponential PDF. Also, distribution of the time between two successive failures of a repairable system is exponential (see Chapter 4).

3.2 Two Main Categories of Reliability Terms

As mentioned above and shown in Table 3.1, the terms used for reliability metrics can be divided into two categories:

1. Reliability metrics
2. Applications of the metrics

The *reliability metrics* are various terms used to quantify the numerical value of the reliability levels, such as mean time between failures (MTBF), failure rate, percent failed, etc.

Table 3.1 Two Main Categories of Reliability Terms

Reliability Metrics	MTBF
	MWBF
	MCBF
	Failures Per Million Hours
	Pr [T> 1000 Hr] = 0.95
	Pr [S] = 0.90
	% Failed
	% Uptime
	% Available
	Down Events per 1000 Cycle
	UM's per Million Cycles
Application of Reliability Metrics	Goals
	Requirements
	Allocations
	Apportionments
	Budgeting
	Inherent Reliability
	Calculated Reliability
	Assessed Reliability
	Predicted Values
	Expected Values
	Warranty

Applications of reliability metrics are terms that are used with any reliability metrics. These terms originate from the reliability related activities. For example, *reliability goal* originates from a goal-setting activity. *Goal* is an application of any reliability metric, e.g., goal MTBF.

3.3 Categories of Reliability Metrics

Reliability metrics can be divided into four main categories:

1. Metrics based on probabilities
2. Metrics based on mean life
3. Metrics normalized by life units
4. Metrics expressed in percentages

3.3.1 Metrics Based on Probabilities

Probabilistic metrics are classical measures of reliability. They always contain a "probability statement." Three typical examples are given below:

1. Probability of performing intended functions for a specified time under stated operational conditions
   Pr [T > 1000 hours] = 0.95
2. Probability of success of a mission
   Pr [S] = 0.95
3. Probability of survival for a specific time under stated conditions
   Pr [T > 1000 hours] = 0.95

3.3.2 Metrics Based on Mean Life

These metrics consist of at least four words, as shown in Figure 3.1. Two of them, *mean* and *between*, are mandatory. Others relate to the measure of life and events.

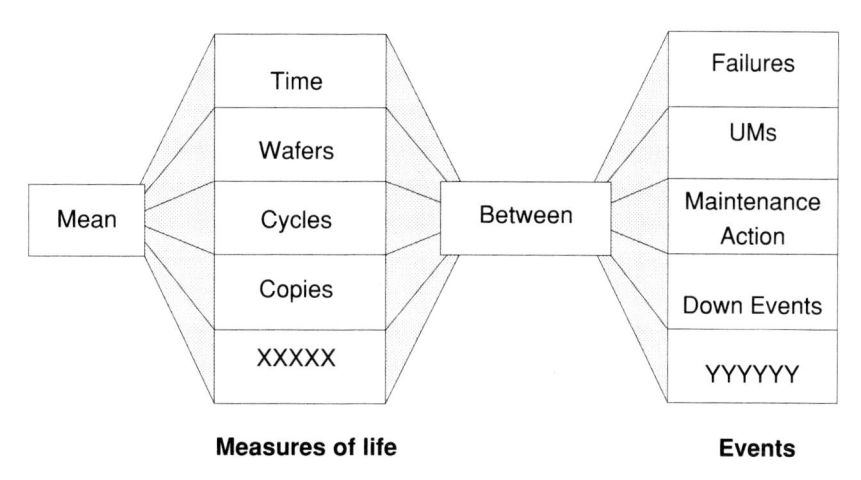

Figure 3.1 Reliability Metrics Based on Mean Life

Using the algorithm given in Figure 3.1, we can make many metrics: Take the word *mean*, select a word for measure of life, take the word *between*, and select the desired event.

EXAMPLES:

- Mean Time Between Failures
- Mean Cycles Between Down Events
- Mean Wafers Between Failures

Mean life is the most widely used category of metric. This category is very widely used to track reliability of semiconductor manufacturing equipment and is recommended by the Semiconductor Equipment Manufacturing International (SEMI), an association of semiconductor manufacturing equipment suppliers, in SEMI Guideline E10-92 for definition and measurement of equipment reliability, availability, and maintainability. See Reference 1.

3.3.3 Metrics Normalized by Life Units

In this category, the numerical values are normalized by a desired number of life units. Metrics in this category also have at least four words. Only one of them is mandatory; the others are selective, as shown in Figure 3.2.

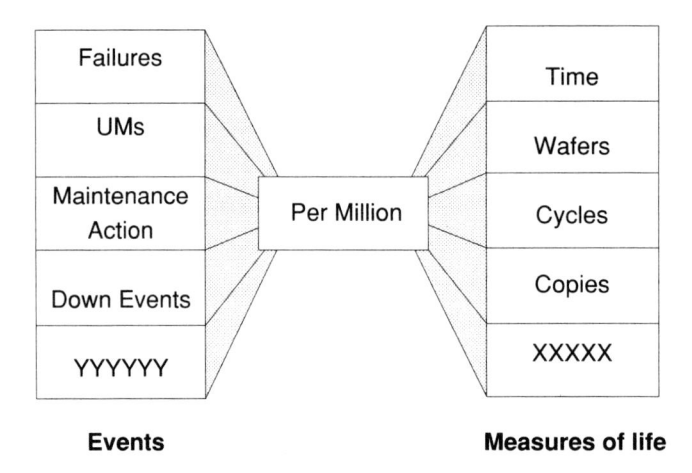

Figure 3.2 Reliability Metrics Normalized by Life Units

Using the algorithm given in Figure 3.2, we can make many metrics of this category: select the desired event, use the word *PER*, select the amount of normalization, and select the measure of life.

EXAMPLES:

- Failures per Million Hours
- Maintenance Actions per Thousand Cycles

3.3.4 Metrics Expressed in Percentages

In this category, metric values are expressed in percentages.

EXAMPLES:

- 2% failed during first 1000 hours/wafers/cycles
- 95% uptime (measure of both reliability and maintainability)
- 90% availability (measure of both reliability and maintainability)

3.4 Applications of Reliability Metrics

The application of reliability metrics can be divided into the following three categories, depending upon the origin of the activities they represent:

1. Desired values
2. Analytical/theoretical values
3. Observed values

3.4.1 Desired Values

In this category, the value of the reliability levels originates from the reliability activities that deal with desires of equipment manufacturers and customers.

FOR EXAMPLE:

The (reliability) goals are what a manufacturer wants his equipment to perform

The (reliability) requirements are what a customer wants his equipment to perform.

Therefore, goals and requirements applications belong to the *Desired Values* category.

When the system-level goals or requirements are broken into department or subsystem level goals or requirements, based on some logical justification, they generate the applications known as allocation, budgeting, or apportionment, which also belong to the Desired Values category.

System-level reliability goals, allocated to subsystem and component, and corresponding operating environments are part of the respective design specifications for reliability. Therefore, design specification is also an application of the reliability metrics and falls in this category.

Figure 3.3 depicts the relationship among various Desired Values applications of the reliability metrics.

3.4.2 Analytical/Theoretical Values

In this category of applications, the value of reliability levels originates from appropriate theoretical reliability activities, such as modeling, part count calculation, and stress analysis.

Following are three typical examples of the Analytical/Theoretical Values applications:

Inherent Reliability: The values are derived from design assessment, assuming benign environments and no error in design, manufacturing, and operation. The inherent reliability is the best achievable level.

Assessed Reliability: When inherent reliability values are adjusted to account for design errors, manufacturing errors and quality problems, human errors, and operating environments, they become assessed values.

Expected/Predicted Reliability: When assessed reliability values are adjusted to account for the planned corrective actions (design modifications), they become expected (or predicted) values.

Figure 3.4 shows the relationship of different applications of the reliability metrics originating from the analytical/theoretical values.

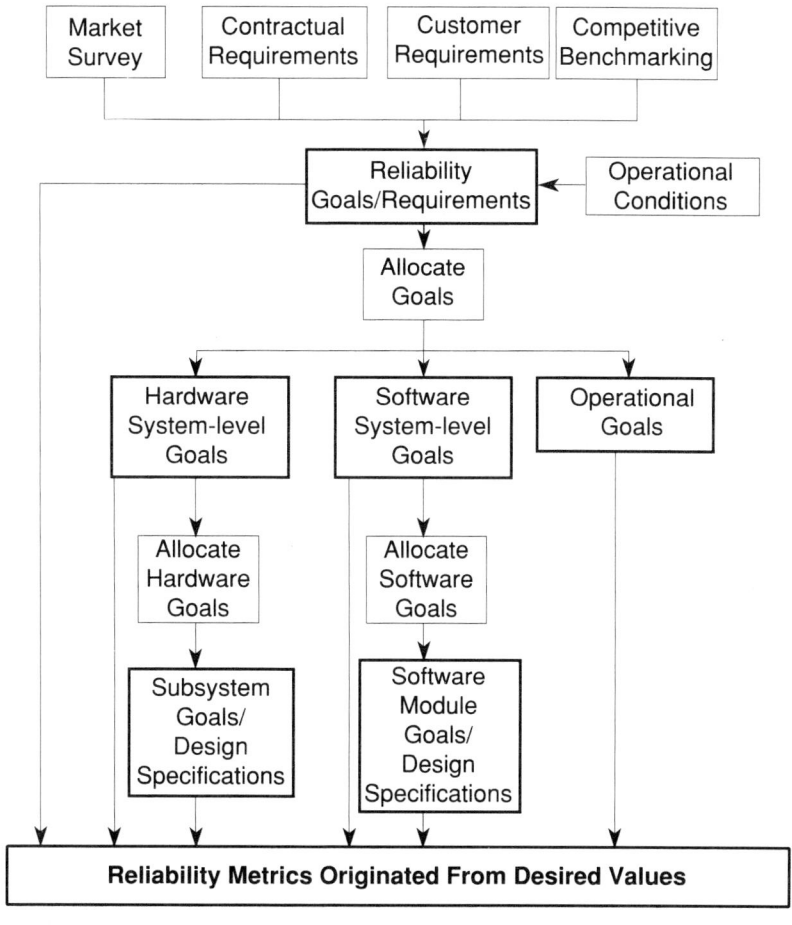

Figure 3.3 Applications of Reliability Metrics Originated From Desired Values

3.4.3 Observed Values

This category represents situations in which the reliability level is established based on actual in-house tests, field tests, or field operations of the equipment.

The following are three typical examples:

Observed Values. The values are derived from actual in-house tests, field tests, or field operations. These values are unaltered as observed values.

```
        ┌──────────────┐
        │    Design    │
        │  Assessment  │
        └──────┬───────┘
               ▼
        ┌──────────────┐
        │   Inherent   │
        │  Reliability │
        └──────┬───────┘
               ▼
     ┌───────────────────────┐
     │     Adjustment for    │
     │ Design and Manufacturing │
     │  Errors and Operational │
     │         Stress        │
     └──────────┬────────────┘
                ▼
         ┌──────────────┐
         │   Assessed   │
         │    Values    │
         └──────┬───────┘
                ▼
          ┌──────────────┐
          │ Adjustment for│
          │    Planned    │
          │Corrective Actions│
          └──────┬───────┘
                 ▼
           ┌──────────────┐
           │  Predicted   │
           │  /Expected   │
           │    Values    │
           └──────┬───────┘
```

Reliability Metrics Originated From Theoretical / Analytical Values

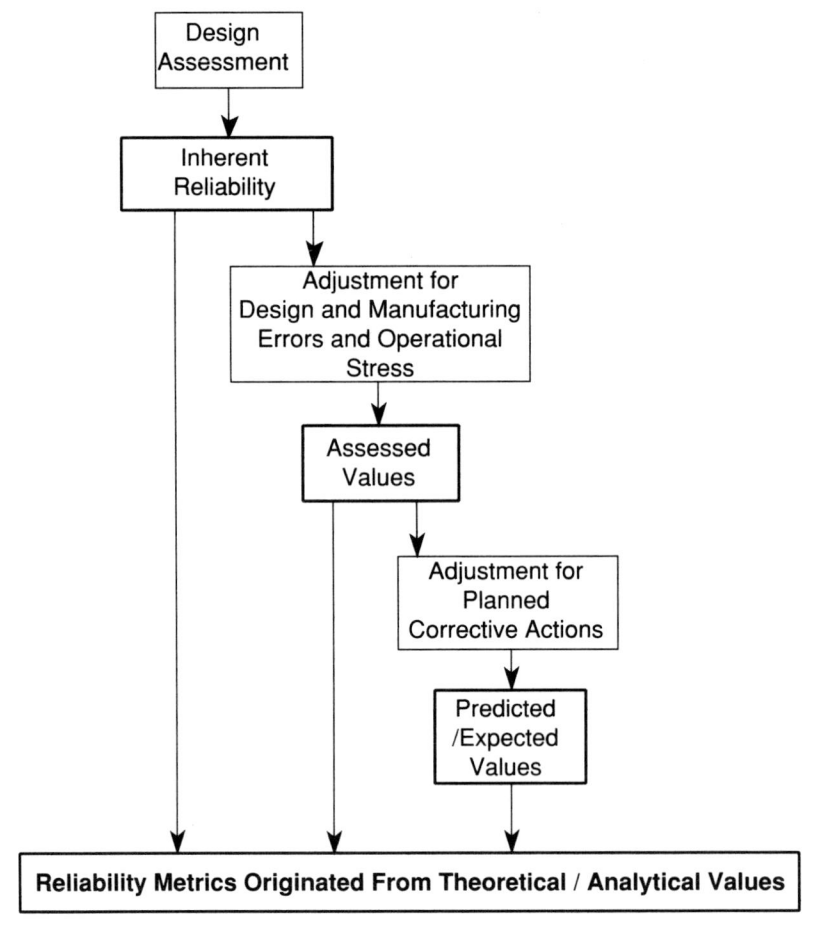

Figure 3.4 Applications of Reliability Metrics Originated From Theoretical/Analytical Values

Assessed Values. When observed values are adjusted to account for non-relevant failures (such as facility problems or out-of-spec consumables) they become assessed values.

Expected/Predicted Values. When assessed reliability values are adjusted to account for planned corrective actions (design modifications), they become expected (or predicted) values.

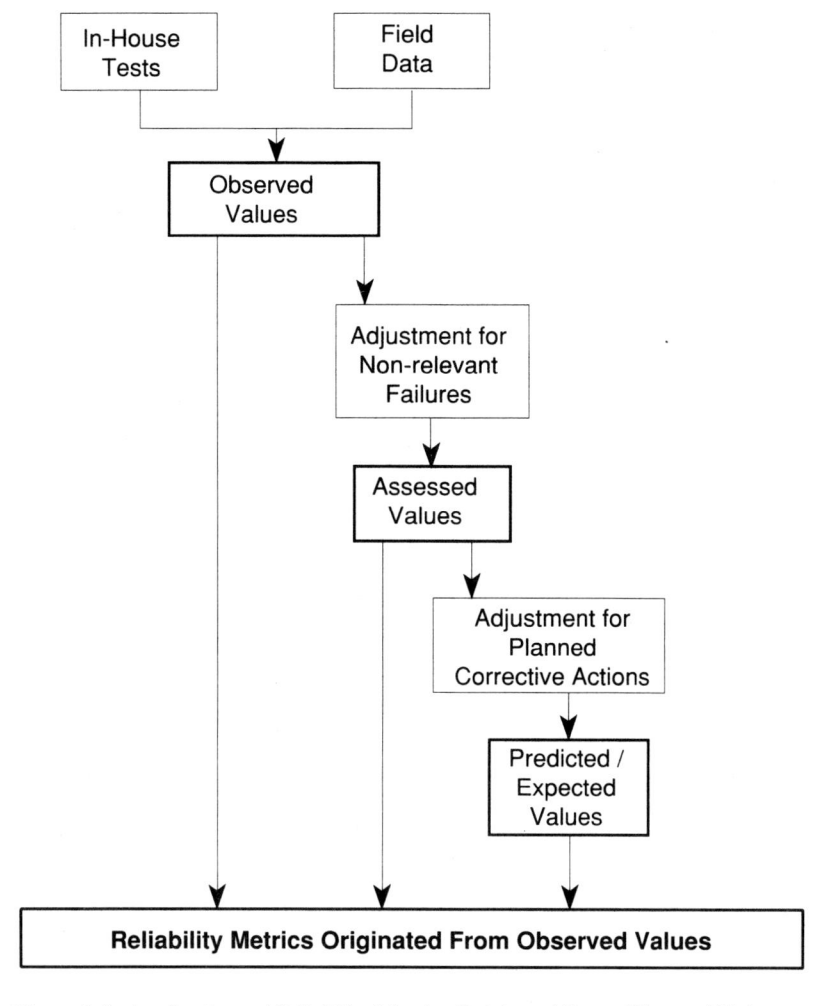

Figure 3.5 Applications of Reliability Metrics Originated From Observed Values

Figure 3.5 shows the relationship of the different applications of the reliability metrics in which the values are based on observed values.

3.5 Precise Use of the Reliability Metrics

Figure 3.6 shows all the most commonly used terms for reliability metrics and their applications. The reliability metrics can be converted from one category to another. For example, if we know MTBF, we can convert it to failures per 1000 hours. Applications of the metrics can neither be converted from one to another nor mixed. However, they can be compared (for example, goal value vs. observed values).

Reliability Metrics			
Probabilistic	Mean Life	Normalized	Percentage
Pr [T > 1000 Hr] = 0.95 Pr [S] = 0.80	MTBF MWBF MCBF	$F/10^6$ Hr UM's/10^6 Cycles	% Failed % Uptime
Reliability Terms			
Goals Requirements Design specifications Allocations Budget Apportionment Warranty	Calculated Inherent Assessed Predicted Expected		Observed Assessed Predicted Expected
Desired	Theoretical		Observed
Applications of Reliability Metrics			

Figure 3.6 Reliability Metrics and Their Applications

We need the following four items to define, fully and precisely, the reliability level of a real-life situation.

1. Appropriate application (e.g., goal)
2. Appropriate reliability metric (e.g., MTBF)
3. Appropriate numerical value (e.g., 5,000)
4. Appropriate unit for the metric (e.g., hours)

This makes an example "Goal MTBF of 5,000 hours."

OTHER EXAMPLES ARE:

- Predicted failure rate of 0.005 failure per 1,000 hours
- Observed mean cycles between failures = 20,000 cycles

When to derive the categories of application and when to use them for a comparison is explained in Chapter 7.

3.6 Standardization of the Reliability Metrics

As we show in this chapter, a multitude of terms exists for reliability metrics and their applications. To focus on the important ones, some industrial sectors have started standardizing definitions for the manufacturing equipment reliability metrics. A typical example is the SEMI E10-92 guideline, Reference 1, developed by SEMI to track reliability performance of the semiconductor manufacturing equipment.

The primary objective of the guideline is to establish a basis for measuring equipment performance in a manufacturing environment. The guideline defines six basic equipment states into which all equipment conditions and periods of time must fall. Figure 3.7 shows these basic states and associated sub-states. The equipment states are determined by functional issues, independent of who performs the function. The equipment reliability metrics concentrate on the relationship of equipment failure to equipment usage (i.e., productive time). MTBF calculations are based on the productive time.

```
                        ┌──────────────┐
                        │ TOTAL TIME   │
                        └──────┬───────┘
                               │
                ┌──────────────┴──────────────┐
                │                             │
         ┌──────┴───────┐                     │
         │ OPERATIONS   │                     │
         │ TIME         │                     │
         └──────┬───────┘                     │
                │                             │
        ┌───────┴────────┐                    │
        │                │                    │
  ┌─────┴─────┐    ┌─────┴──────┐             │
  │ EQUIPMENT │    │ EQUIPMENT  │             │
  │ UPTIME    │    │ DOWNTIME   │             │
  └───────────┘    └────────────┘             │
```

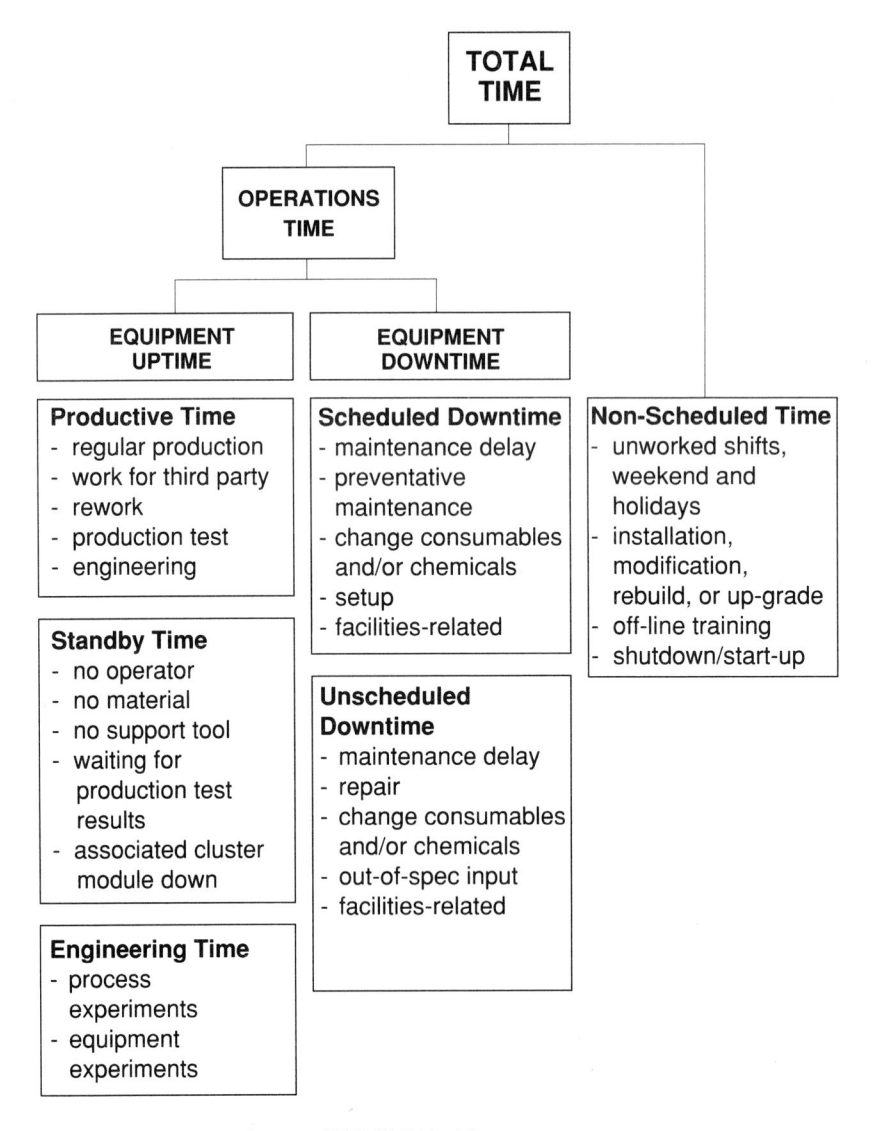

Productive Time
- regular production
- work for third party
- rework
- production test
- engineering

Standby Time
- no operator
- no material
- no support tool
- waiting for production test results
- associated cluster module down

Engineering Time
- process experiments
- equipment experiments

Scheduled Downtime
- maintenance delay
- preventative maintenance
- change consumables and/or chemicals
- setup
- facilities-related

Unscheduled Downtime
- maintenance delay
- repair
- change consumables and/or chemicals
- out-of-spec input
- facilities-related

Non-Scheduled Time
- unworked shifts, weekend and holidays
- installation, modification, rebuild, or up-grade
- off-line training
- shutdown/start-up

Figure 3.7 Equipment States of SEMI E10–92

This guideline also provides a formula for calculating MTTR, uptime, and utilization. Uptime has three levels: (i) equipment dependent, (ii) supplier dependent, and (iii) operational. Utilization has two levels: (i) operational, and (ii) total. A summary of key formulas of SEMI E10-92 are shown in Figure 3.8.

$$MTBF_P = \frac{PT}{N} \quad (1)$$

$$MCBF = \frac{\text{Total Equipment Cycles}}{N} \quad (2)$$

$$\text{Equipment-Dependent Uptime} = \frac{UPT}{OT - AMDT - OSDT - FDT} \times 100 \quad (3)$$

$$\text{Supplier-Dependent Uptime} = \frac{UPT}{OT - UMDT - OSDT - FDT} \times 100 \quad (4)$$

$$\text{Operational Uptime} = \frac{UPT}{OT} \times 100 \quad (5)$$

$$\text{Operational Utilization} = \frac{PT}{OT} \times 100 \quad (6)$$

$$\text{Total Utilization} = \frac{PT}{TT} \times 100 \quad (7)$$

Where:
- $MTBF_P$ = Mean Productive Time Between Failures
- MCBF = Mean Cycles Between Failures
- PT = Productive Time
- N = Number of Failures
- UPT = Equipment Uptime
- OT = Operations Time
- AMDT = All Maintenance Time
- OSDT = Out-of-Spec Input Down Time
- FDT = Facility Related Down Time
- UMDT = User Maintenance Delays
- TT = Total Time

Figure 3.8 Key Formulas of SEMI E10–92

SEMI E10-92 is very widely used in semiconductor manufacturing. Eventually, this guideline will become a standard for tracking performance of semiconductor manufacturing equipment and a benchmark for other industries.

REFERENCE

1. SEMI E10-92, *Guideline for Definition and Measurement of Equipment Reliability, Availability, and Maintainability (RAM)*, SEMI International Standard, Equipment Automation/Hardware, Mountain View, CA, 1992.

Chapter 4
Reliability of Systems

We defined *component* and *system* in Chapter 2, Section 2.8. In this chapter, we will present the analytical/theoretical relationship between the reliability level of components and systems. A system's reliability level depends upon the reliability of its components and how well they are functionally related to the system.

First, we will look at the following three basic systems:

1. Series system
2. Parallel system
3. Standby system

In the final section of this chapter, we will expand on the repairable system (defined in Chapter 2) and determine the distribution of time between two successive failures.

4.1 Series System

The series system is probably the most common configuration for modeling equipment. It is also the simplest to analyze. In a series system, all components must perform their intended function successfully if the system is to perform its intended function (i.e., if any component fails, the entire system fails).

The block diagram of a series system is given in Figure 4.1. Mathematically, the reliability level of a series is given by

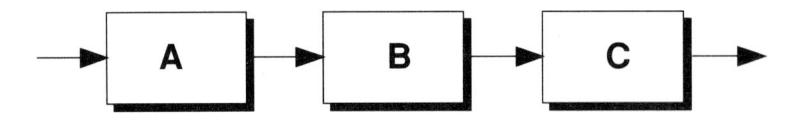

Figure 4.1 Series System

$$R_s = R_1 \times R_2 \times R_3 \times \ldots \ldots \ldots \times R_n \qquad (4.1)$$

WHERE:

R_s = System-level reliability

R_n = Reliability level of nth component

For the exponential distribution, system MTBF is given by

$$MTBF_s = \frac{1}{\frac{1}{MTBF_1} + \frac{1}{MTBF_2} + \frac{1}{MTBF_3} + \ldots + \frac{1}{MTBF_n}} \qquad (4.2)$$

WHERE:

$MTBF_s$ = System-level MTBF

$MTBF_n$ = MTBF of nth component

For example, for a three component series system, if $MTBF_1$ = 5000 hours, $MTBF_2$ = 4000 hours, and $MTBF_3$ = 2000 hours, then

$MTBF_s$ = 1052 hours.

Note that the system-level MTBF is shorter than that of the component with the shortest MTBF.

4.2 Parallel System

A parallel system is one that fails only when all the components in parallel fail. Parallel arrangement of components is usually used to improve system reliability.

The block diagram of a parallel system is given in Figure 4.2. Mathematically, the reliability level of a parallel system is given by:

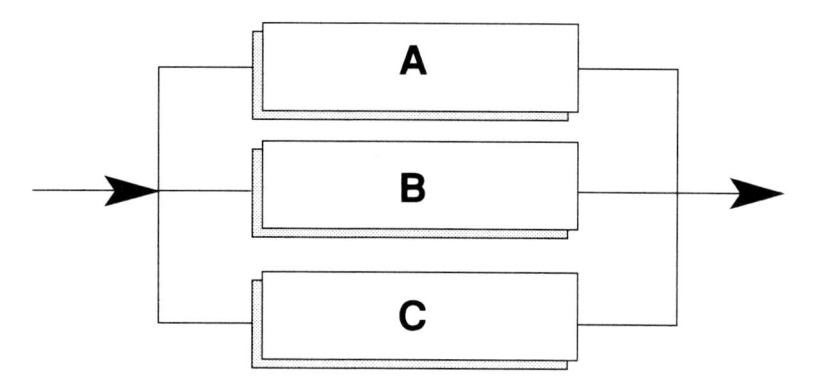

Figure 4.2 Parallel System

$$R_s = 1 - \{(1-R_1) \times (1-R_2) \times (1-R_3) \times \ldots \times (1-R_n)\} \qquad (4.3)$$

WHERE:

R_s = System-level reliability

R_n = Reliability level of nth component

For a system with two parallel components, with the exponential PDF, the system MTBF is given by:

$$MTBF_s = \frac{1}{\lambda_1} + \frac{1}{\lambda_2} - \frac{1}{\lambda_1 + \lambda_2} \qquad (4.4)$$

WHERE:

λ_1 and λ_2 = Failure rates of components 1 and 2, respectively

For example, if component 1's MTBF = 1000 hours, and component 2's MTBF = 1000 hours, then

$MTBF_s$ = 1500 hours

Note that system-level MTBF of a parallel system is larger than that of any individual component.

As shown in Reference 1, equation (4.4) becomes very lengthy as the number of components increases.

4.3 Standby System

The parallel system introduced in the previous section is a "pure" parallel system. In real life, variations of the parallel system are more typical. One of the most common variations is the standby system which uses an active component, idle component(s) that only operates when necessary, a sensing device, and a switching device. When the sensing device detects failure of the active component, the switching device switches a load to an idle component.

In contrast, the pure parallel system operates all of its components simultaneously.

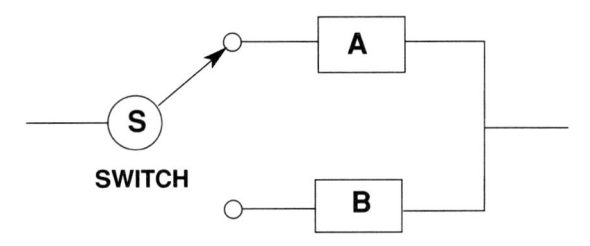

Figure 4.3 Standby System

Figure 4.3 shows a typical two-component standby system. The system-level reliability depends upon the sensing and switching mechanisms and the component repair policy. The following example situations illustrate the relationship between component and system-level reliability for standby systems. Both examples assumes that the sensing mechanism is capable of sensing a failure instantaneously and that the switching mechanism switches the load to the idle component instantaneously.

1. If repair policy calls for instantaneous repair of the failed component, then the standby system will never fail.
2. Repair policy is to repair only after both components fail. Then the standby system MTBF is given by

$$MTBF_s = MTBF_1 + MTBF_2 \qquad (4.5)$$

Note that two components with MTBF = 1000 hours in a standby system yield a system-level MTBF = 2000 hours.

4.4 Repairable System

As defined in Chapter 2, a system that, after failing to perform at least one of its intended functions, can be restored to perform all of its intended functions by any method other than replacing the entire system is called a *repairable system*. The repairable system can be restored by replacing, repairing, adjusting, or cleaning the appropriate component(s). Most large systems such as semiconductor manufacturing equipment are repairable systems.

In a repairable system, the distribution of failure times between two successive failures is of prime interest. If we assume that:

1. Each component failure is an independent renewal process, i.e., when a component fails, it is replaced by a new component and this does not affect any other components
2. The system is a series system with many independent components.

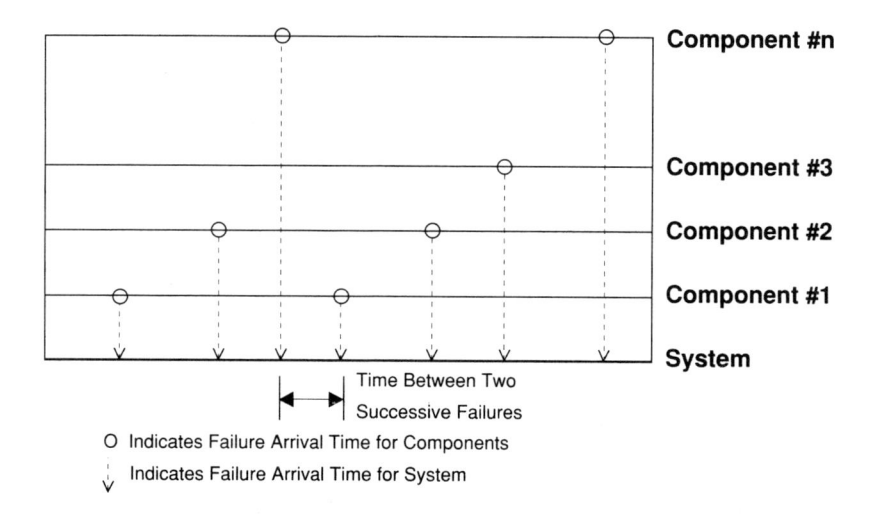

Figure 4.4 Repairable System as a Superimposed Renewal Process

Then under very generic conditions, the system-level failure is a superimposed renewal process as shown in Figure 4.4. The time between two successive failures will approximately follow an exponential distribution (See References 2 and 3 for more details). This approximation makes reliability analysis of a repairable system very easy. We need to know only one MTBF parameter of the distribution to be able to perform system-level reliability analyses.

REFERENCES

1. Dimitri Kececioglu, *Reliability Engineering Handbook*, Volume 2, PTR Prentice Hall, Englewood Cliffs, NJ, 1991.

2. Paul A. Tobias and David C. Trindale, *Applied Reliability*, Van Nostrand Reinhold, New York, NY, 1994.

3. Harold Ascher and Harry Feingold, *Repairable Systems Reliability*, Marcel Dekker, Inc., New York, NY, 1984.

Chapter 5
Reliability Compared With Other Disciplines

Reliability is a widely recognized and used discipline in developing, manufacturing, and servicing military and consumer products. Because of its broad use, it relates to many other disciplines within industry. In this chapter, we will compare reliability with related disciplines and identify similarities and differences.

5.1 Quality and Reliability

There is always a point of contention between quality advocates and reliability advocates. Is reliability part of quality or quality part of reliability? Experts have argued both ways for a long time. A simple answer is that if reliability is a customer requirement then reliability is a part of quality—one of the quality characteristics. However, quality definition has no time base. When you stretch the quality characteristic of performing intended functions forward in time then it becomes reliability as shown in Figure 5.1. In this regard, quality is a one-dimensional characteristic, while reliability is a two-dimensional characteristic.

Quality relates to goodness of an item at a given instant. Most measures of quality are qualitative. Reliability relates to life longevity. Most of the reliability metrics (see Chapter 3) are quantitative.

5.2 Quality Control, Quality, and Reliability

Let us first understand the difference between quality control (QC) and quality. *Quality* is a characteristic and *QC* is a process or method that helps to achieve and control the desired level of the quality characteristic.

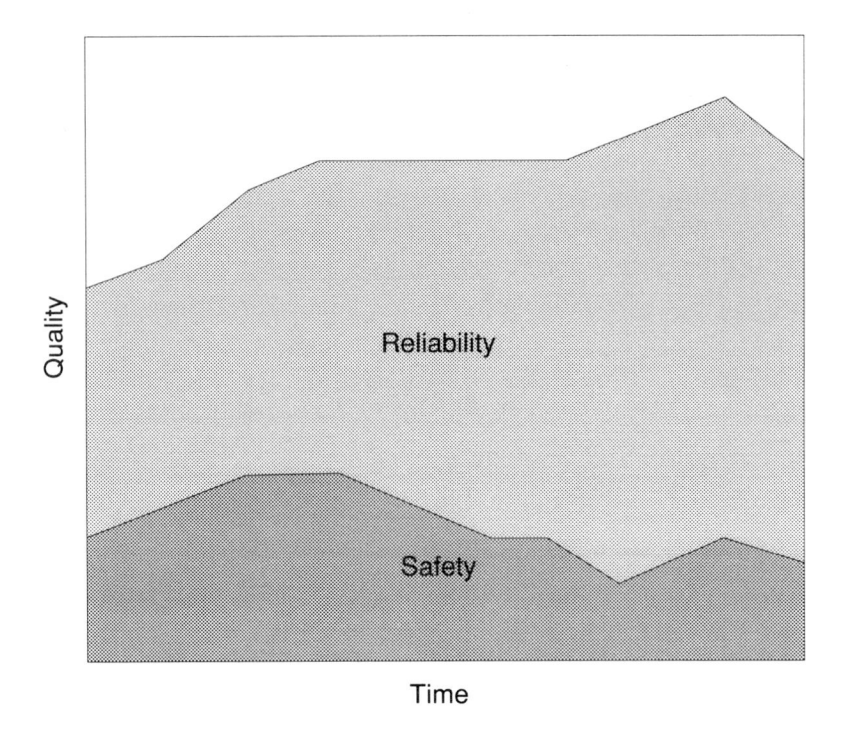

Figure 5.1 Relationship Among Quality, Reliability, and Safety

QC assures that equipment designs (drawings) are converted to parts, components, subsystems, equipment, and systems, according to the specifications and without degrading the designed-in reliability. QC also assures that the manufacturing processes result in uniform equipment within the specified limits. In short, QC assures design and manufacturing quality.

Reliability assures that the proper design concepts and methodologies are used to achieve the desired reliability level.

Another contrasting point is that QC deals with the defective percentage, mostly in the manufacturing stage, at a given point in time. Reliability deals with the behavior of failure rates throughout all periods of the equipment existence.

5.3 Safety and Reliability

Safety, as a characteristic of equipment, refers to being in a safe state and creating safe environments during manufacturing, installation, operation, and maintenance. This means there is no harm to anyone in any way (short term or long term) who is working with or in the vicinity of the equipment being manufactured, installed, operated, or maintained.

There is much similarity between reliability and safety. Both require a proactive approach and should be designed-in to achieve the desired level. Both deal with the equipment over all periods of the equipment existence. Although every reliability failure may not affect safety, any safety violation situation during the operation of equipment is considered as a failure within a reliability calculation. In this regard, safety is a subset of reliability, as shown in Figure 5.1.

Since we now understand the safety factor, we can modify the formal definition of reliability (Chapter 2) by adding a qualifier to read "performing intended functions for a specified time under the stated conditions, safely." This modification links safety and reliability.

5.4 Maintainability and Reliability

There is no direct relationship between reliability and maintainability. Reliability deals with the operational life longevity and failures of equipment, while maintainability deals with restoring the equipment operation and the time it takes to restore it. However, both disciplines are complementary and both support high-level equipment performance metrics such as availability and OEE (see Chapter 6).

Formally, *maintainability* is the probability that the equipment will be restored to a specific operational condition (able to perform its intended functions) within a specified period of time, when the maintenance is performed by personnel having specified skill levels and using prescribed procedures, resources, and tools. Maintenance can be either unscheduled or scheduled. One of the most popular measures of maintainability is Mean Time To Repair (MTTR).

5.5 Availability and Reliability

Availability is a joint measure of reliability and maintainability. It is defined as the probability that the equipment will be in a condition to perform its intended functions when required. One popular formula for calculating availability is

$$\text{Availability} = \text{MTBF}/(\text{MTBF} + \text{MTTR}) - (\% \text{ PM Time})/100 \tag{5.1}$$

One of the most widely used parameters of the availability is % Uptime, which is defined as

$$\%\text{Uptime} = \text{Availability} \times 100 \tag{5.2}$$

As evident by the above formulas, uptime (availability) is influenced both by reliability and by maintainability. Figure 5.2 shows a few typical graphs depicting this relationship. If reliability is high (high MTBF) then maintainability has very little effect on the availability. Maintainability and reliability have significant affect on availability when MTBF is low and MTTR is high.

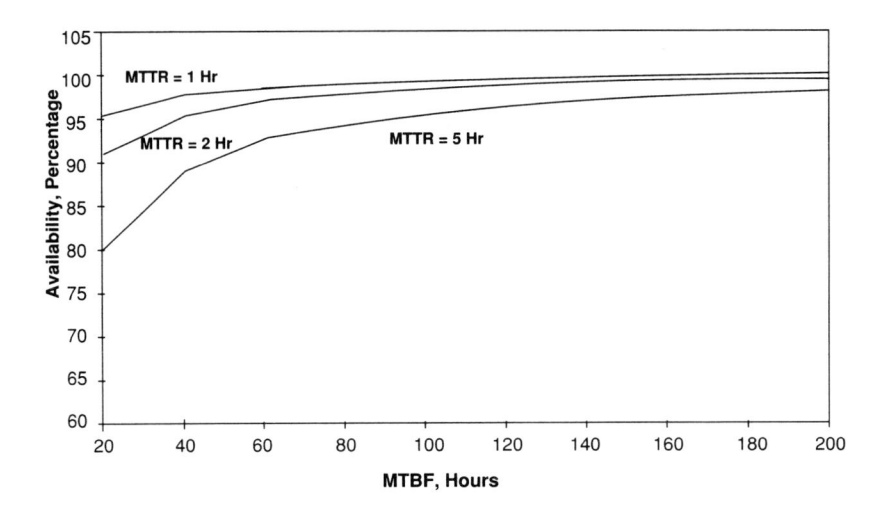

Figure 5.2 Availability vs. MTBF

Besides availability, there are other high-level equipment performance metrics that are related to and based on reliability. These metrics are presented in Chapter 6.

5.6 Which Is More Important?

Which is more important, quality, reliability, safety, maintainability, or availability? The answer to this question is not a simple one. It depends on the situation. For example:

If defect-free equipment deliveries and consistent operations are necessary, then quality and QC are important.

If the consequences of a failure are very high—either very costly to repair, lots of scraped product units, or harmful to human beings—reliability and safety are most important. These situations demand an MTBF as large as possible.

If it is important to have equipment available whenever we want it, availability is most necessary. These situations require high MTBF and low MTTR.

Chapter 6
Reliability in the High-Level Equipment Performance Metrics

In real life, we always face a question, "What should be the proper level of reliability?" A simple answer would be "the higher, the better." However, the answer cannot be that simple. To answer this question, our industry has developed high-level equipment performance metrics for which reliability is a key element. The proper level of reliability is the one that yields the optimal value of the high-level metric being considered.

These high-level equipment performance metrics are becoming increasingly important to compete in the global market because they satisfy customers' reliability requirements in an optimum manner.

We already defined one high-level metric, availability, in Chapter 5, Section 5.5. We also studied the effect of reliability on availability. In this chapter, we will describe the following three high-level equipment performance metrics:

1. Overall Equipment Effectiveness (OEE)
2. Life Cycle Cost (LCC)
3. Cost of Ownership (CoO)

6.1 Overall Equipment Effectiveness

Overall Equipment Effectiveness is the most recent high-level equipment performance metric. It was developed as an equipment effectiveness metric in Japan to measure the effectiveness of a manufacturing technique called Total Productive Maintenance (TPM). The American

Institute for Total Productive Maintenance (AITPM) is currently the major sponsor of using the OEE metric in the U.S. Gradually, OEE use is increasing in the U.S.A.

OEE is an all-inclusive metric of equipment productivity, i.e., it is based on reliability (MTBF), maintainability (MTTR), throughput, utilization, and yield. All the above factors are grouped into the following three sub-metrics of equipment effectiveness.

1. Availability
2. Performance efficiency
3. Rate of quality

The three sub-metrics and OEE are mathematically related as follows:

$$\text{OEE\%} = \text{Availability} \times \text{Performance} \times \text{Rate of Quality} \times 100 \qquad (6.1)$$

Now let us look at each OEE sub-metric in more detail.

6.1.1 Availability

We have already defined availability in Chapter 5. There are many variations in the availability calculation, most of which stem from their definitions of productive time. In this book, we will stick to the following simple definition, but acknowledge that more precise calculations are available.

$$\text{Availability} = \frac{TT - NST - SMT - USMT}{TT} \qquad (6.2)$$

WHERE:

TT = Total Calendar Time
NST = Nonscheduled Time (such as holidays or weekends)
SMT = Scheduled Maintenance Time
USMT = Unscheduled Maintenance Time

6.1.2 Performance Efficiency

The performance efficiency is based on the losses incurred from idling, minor stops, and equipment speed losses. It is given by:

$$\text{Performance Efficiency} = \frac{\text{Theoretical CT}}{\text{Actual CT}} = \frac{\text{Actual PPH}}{\text{Theoretical PPH}} \qquad (6.3)$$

WHERE:

CT = Cycle Time
PPH = Throughput Rate in Parts Per Hour

6.1.3 Quality Rate

The quality rate is a measure of output quality and it is given by:

$$\text{Quality Rate} = \frac{\text{Total Part Produced} - \text{Rejects}}{\text{Total Parts Produced}} \qquad (6.4)$$

6.1.4 An Example of OEE Calculations

The following facts are known. In a period of one week, a piece of equipment

1. was not scheduled for production for 48 hours
2. was down for scheduled and unscheduled maintenance for 2 and 4 hours, respectively
3. had production rate of 80 PPH vs. its theoretical PPH of 100
4. yielded 15 rejects out of the total 9120 production units

Using equations (6.1) through (6.4), we have

$$\text{OEE\%} = \frac{168 - 48 - 2 - 4}{168} \times \frac{80}{100} \times \frac{9120 - 15}{9120} \times 100 \qquad (6.5)$$

or OEE% = $0.678 \times 0.8 \times 0.998 \times 100 = 54.1\%$.

Figure 6.1 depicts the relationship between reliability (MTBF) and OEE.

Note that OEE does not directly include any cost-related factor in its calculations.

6.2 Life Cycle Cost

Life Cycle Cost (LCC) is the total cost of acquiring and operating an equipment over its entire life span. LCC includes all supplier and customer costs incurred from the point at which the decision is made to acquire the equipment, through operational life, to eventual disposal of the equipment.

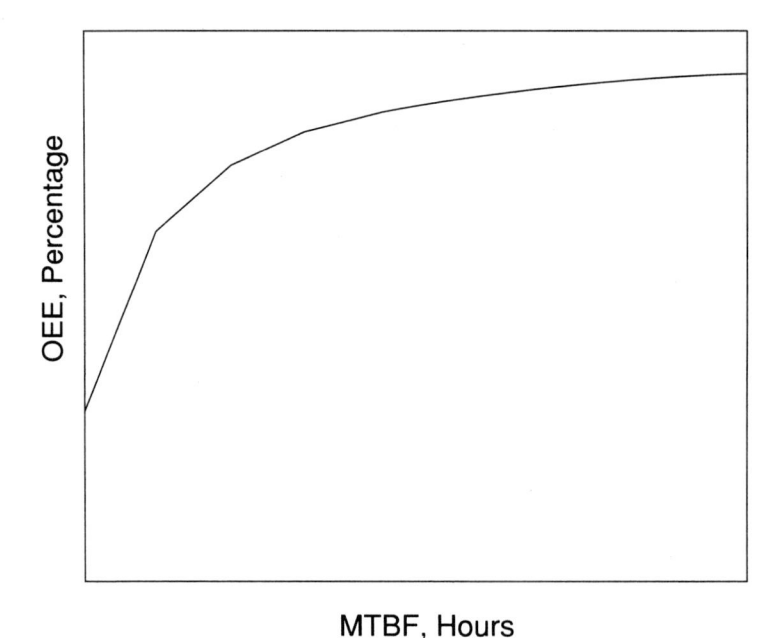

Figure 6.1 A Typical MTBF vs. OEE Graph

LCC is an older metric than OEE and is based on the equipment's cost factors. It has been in use for many years to determine the optimal level of reliability that generates minimum LCC, as shown in Figure 6.2. It has also been used to perform trade-off between acquisition and operational costs.

Two main elements of the LCC are: (1) acquisition cost and (2) operational cost. Supplier costs plus the supplier's gross profit are referred to as *acquisition costs* that are passed to the customer in the purchase price of the equipment. These costs include:

- Research and development
- Basic engineering and design
- Testing and evaluation
- Manufacturing, parts, materials, tools, and labor
- Quality and reliability assurance

- Supplier-provided packaging, shipping, and installation
- Supplier-provided training and support

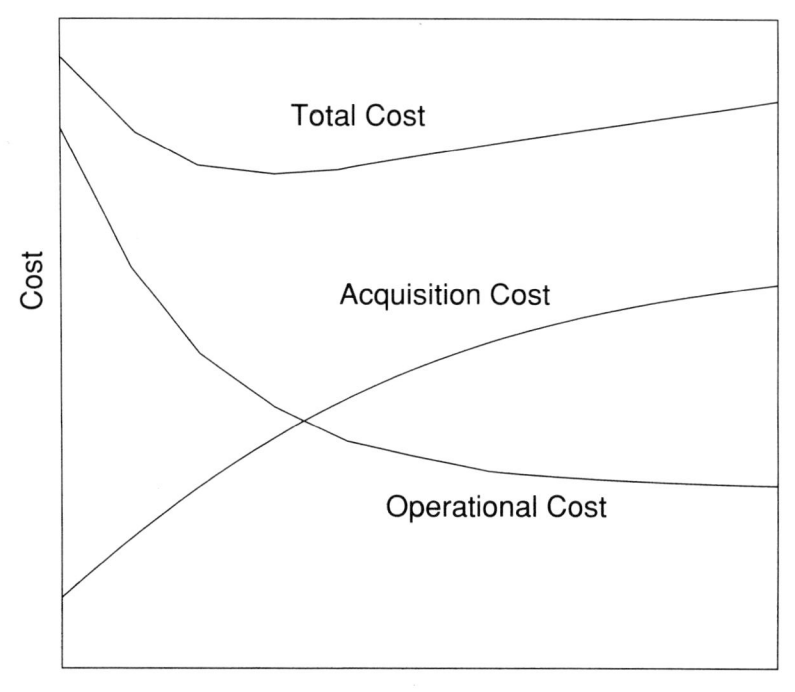

Figure 6.2 A Typical Reliability Level vs. Life Cycle Cost

- Marketing and sales
- Warranty costs
- Supplier profit

Costs incurred by the customer are referred to as operational costs and include:

- Supply process management
- Customer installation and training
- Repair and maintenance (unscheduled and scheduled) costs

- Spare parts and their inventory
- Operational personnel
- Training
- Facility and utilities
- Materials
- Cost of scraping the equipment less the scrap value

Since all elements of LCC do not occur in the same year, we need to consider one more factor called "time value of money" in the LCC calculations. This factor converts all costs incurred after the first year to an equivalent present value (worth) in the first year. The present value discount factors are found in any interest rate table book. Table 6.1 contains an example of typical LCC calculations.

Table 6.1 A Typical Life Cycle Cost Calculation

	Year in Which Cost Incurred					
	1995	1996	1997	1998	1999	2000
Acquisition Cost	500,000					
Installation Cost	50,000					
Operational Cost	35,000	50,000	55,000	60,000	65,000	70,000
Repair and Maintenance Cost	30,000	33,000	36,000	39,000	44,000	50,000
Removal Cost						10,000
Scrap Value						(5,000)
Subtotals*	615,000	83,000	91,000	99,000	109,000	125,000
Present Worth Factor*	1.0000	0.9259	0.8573	0.7938	0.7350	0.6806
Present Worth in 1995 Dollars	615,000	76,852	78,018	78,589	78,589	85,073
	Total LCC in 1995 Dollars = 1,013,650					

* Assuming 8% rate of interest (inflation rate)

A variety of LCC models are available. Reference 1 contains the most commonly used LCC models.

6.3 Cost of Ownership

LCC is one of the most widely used high-level metrics, but it has the following shortcomings. It does not include:

- Effect of the production volume
- Product scrap loss because of poor quality output
- Consumable cost
- Waste disposal cost
- Taxes, insurance, and interest expenses

To overcome the above, SEMATECH developed a Cost of Ownership (CoO) model, as described in Reference 2, which calculates the true cost of ownership per good unit produced in a given time period, usually a calendar year.

The CoO depends upon the production throughput rate, equipment acquisition cost, equipment reliability, throughput yield, and equipment utilization. The basic CoO is given by the following equation.

$$\text{CoO per unit} = \frac{FC + OC + YLC}{P \times THP \times Y \times U} \qquad (6.6)$$

WHERE:

FC = Fixed costs (amortized for the period under consideration)
OC = Operating costs
YLC = Yield loss costs
P = Time period
THP = Throughput rate
Y = Throughput yield
U = Utilization

Let us define the above factors in a little more detail.

Fixed Costs: The fixed costs are typically determined from a variety of items such as: purchase price, taxes and duties, transportation costs, installation cost, start-up cost, and training cost. The allocation to the time period under consideration is strictly a function of allowable depreciation schedule and the length of the time period.

Operating Costs: The operating costs for a piece of equipment are consumable and material, maintenance, parts, waste disposal, and operators.

Yield Loss Costs: The yield loss costs are those associated with lost production units that are directly attributable to equipment performance.

Throughput Rate: Throughput rate is the production rate of the equipment, usually expressed in parts per hour (PPH).

Throughput Yield: Throughput yield is the fraction of good units produced. It is determined by

$$Y = \frac{\text{Total Units Produced} - \text{Defective Units Produced}}{\text{Total Units Produced}} \qquad (6.7)$$

Utilization: Utilization is the fraction of the total time that a tool is available for production. It is determined by

$$U = \frac{TT - USMT - SMT - SBT - ET}{TT} \qquad (6.8)$$

WHERE:

TT = Total time
USMT = Unscheduled maintenance time
SMT = Scheduled maintenance time
SBT = Standby time
ET = Engineering time (including re-qualification)

Table 6.2 contains an example of a simple CoO calculation. For more elaborate CoO calculations, refer to Reference 2.

Table 6.2 A Typical Cost of Ownership Calculation

CoO Input Data:	
Equipment Acquisition Cost = $ 1,000,000	Equipment Life = 5 Years
Throughput Rate = 20 Units/Hour	Throughput Yield = 0.98
Operation Cost = $ 800,000/Year in 1995	Part Cost = $ 50,000/Year in 1995
Labor Rate = $ 50/Hour in 1995	PM Time = 10 Hours per Month
MTBF = 200 Hours	MTTR = 2 Hours
Utilization - 75%	Inflation Rate = 4% per Year

CoO Calculations					
	Year				
Cost Factors	1995	1996	1997	1998	1999
Depreciation, $	200,000	200,000	200,000	200,000	200,000
Operational cost, $	800,000	832,000	865,280	899,891	935,887
Repair & Main. Cost, $	66,320	68,973	71,132	74,601	77,585
Yield Loss, $	250,000	260,000	270,400	281,216	292,465
Total Cost, $	1,116,320	1,160,973	1,207,412	1,255,708	1,305,937
Good Unit Produced	128,772	128,772	128,772	128,772	128,772
CoO per Unit, $	8.67	9.02	9.38	9.75	10.14

Many functional areas use the CoO value as one of their critical factors, including competitive analysis, benchmarks analysis, materials cost impact analysis, equipment selection, and project prioritizing. In addition, CoO analysis is an excellent marketing tool for sales personnel.

6.4 Hierarchy of Equipment Performance Metrics

Figure 6.3 depicts the hierarchy of equipment performance metrics. As shown in the figure, when we add time dimension to quality and safety, it becomes reliability. Reliability and maintainability jointly make up availability. When production speed efficiency and production defect rate are combined with availability, it becomes Overall Equipment Effectiveness (OEE). Acquisition and operational cost make up Life Cycle Cost (LCC). When scrap, waste, consumables, tax, and insurance cost are added to LCC and the total is normalized by the production volume, it becomes Cost of Ownership (CoO).

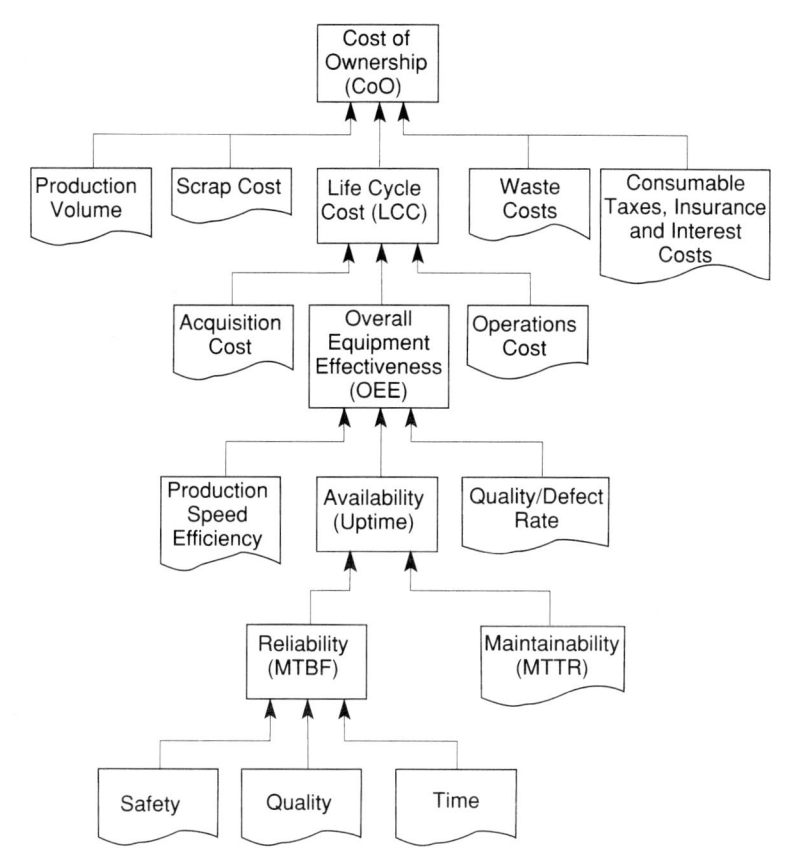

Figure 6.3 Hierarchy of High-Level Equipment Performance Metrics

REFERENCES

1. MIL-HDBK-338, *Electronic Reliability Design Handbook*, 15 October 1984.

2. SEMATECH, *Cost of Ownership Model*, Technology Transfer Document # 91020473B-GEN, SEMATECH, Inc., Austin, TX, 1992.

Chapter 7
Equipment and Equipment Program Life Cycle Phases

Before we describe any life cycle phases, let us first understand the difference between *equipment* and *equipment program*. Equipment refers to an individual unit of equipment with a unique serial number that is used by a specific customer. It may be identical to other units, but it has a unique serial number.

On the other hand, an equipment program represents all the activities from concept of the equipment line, to the discontinuation of the equipment line. This includes design, prototype testing, and production of an entire generation of one equipment line. In a specific equipment program, many individual units of equipment are manufactured and sold. Also, the basic equipment design during an equipment program remains almost unchanged, although it progressively improves from first release to last.

For example: A customer buys a plasma etcher model EZ2, serial number 112 from an equipment supplier. For the customer, the plasma etcher serial number 112 is *equipment*. For the supplier, it is a unit of the plasma etcher model EZ2 *equipment program*.

Now let us understand life cycle phases of the equipment and equipment program. A working knowledge of these phases enables proper planning and execution of the activities and functions necessary for design-in, build-in, and management of reliability growth in a cost effective manner.

7.1 Equipment Life Cycle Phases

The life of typical equipment has three phases:

Early Life: This phase is also known as the infant mortality phase. Failure rate during this phase decreases with equipment age because the rate of manufacturing workmanship errors, insufficient burning-in or screening, and improper debugging after the final assembly reduces over time and increased experience.

Useful Life: This phase is also known as the constant failure rate phase. The failure rate remains at a constant level after decreasing from a high level during the early life phase. All the failures in this phase are categorized as random or chance failures.

Wear-out Life: During this phase, the failure rate starts going up again, because critical parts are wearing out.

These three life cycle phases are shown in Figure 2.8.

7.2 Equipment Program Life Cycle Phases

An equipment program life cycle begins when the idea for the equipment design is conceived and ends when the equipment is no longer manufactured. The equipment program life cycle is divided into the following six phases, as shown in Figure 7.1:

1. Concept and feasibility phase
2. Design phase
3. Prototype phase
4. Pilot production phase
5. Production phase
6. Phase out phase

Figure 7.1 shows two different formats of the above phases: (i) in sequential format, and (ii) concurrent format. Most modern organizations prefer the concurrent format.

Let us examine these phases in detail.

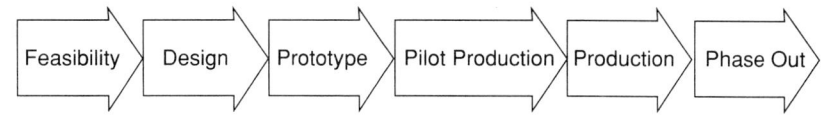

Equipment Life Cycle Phases in Sequential Format

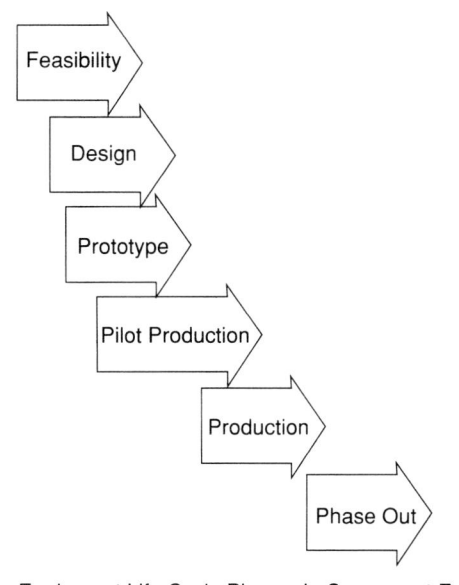

Equipment Life Cycle Phases in Concurrent Format

Figure 7.1 Equipment Program Life Cycle Phases in Sequential and Concurrent Formats

7.2.1 Concept and Feasibility Phase

The life cycle begins here. The need for new equipment is identified and alternative approaches to fulfilling that need are explored. The need could be driven by existing equipment that can no longer performs its intended functions or by customer requirements that no existing equipment can provide.

During this phase, marketing and sales personnel, customer service representatives, design and reliability engineers, and manufacturing engineers work with the customer to:

- Determine the need for new equipment
- Establish reliability goals
- Evaluate the feasibility of meeting these goals
- Examine alternative design concepts
- Perform preliminary cost tradeoffs

7.2.2 Design Phase

The alternative design concepts, selected during the concept and feasibility phase, are explored in more detail by the design engineers during this phase of the life cycle. Reliability and manufacturing engineers, as well as quality assurance and field service personnel, are generally called on by the design engineers for input regarding parts selection, serviceability, and manufacturing processes. Also, reliability goals set for the equipment during the concept and feasibility phase are translated into requirements very early in the design phase.

Most of the designing-in reliability activities are performed during this phase. These activities include:

- Simplification of equipment design
- Derating
- Use of proven components and methods
- Redundancy
- Design reviews

Chapter 9 elaborates the above activities.

Several iterations of design review and redesign are required before a design is ready for prototype construction. Design reviews are important for measuring the progress of design requirements and for gaining management approval to proceed with the prototype phase of the life cycle. These reviews are conducted in parallel with the design process.

7.2.3 Prototype Phase

Specific designs that are selected during the design phase are built and tested to determine if all design requirements are met. The prototype phase provides the first opportunity to validate the entire design and is commonly called alpha-site evaluation. Selected customers are included in alpha-site evaluations and are asked to provide feedback on all aspects of the equipment. Multiple design alternatives are required for prototyping and testing whenever a serious question exists concerning the best overall choice.

Design reviews are continued to give the customer an opportunity to review the latest design being considered.

Concurrent with redesigns and design reviews, reliability engineers, quality assurance personnel, and manufacturing engineers develop quality assurance plans, design inspection and testing programs, set up production facilities, and develop production plans in preparation for the pilot production phase.

7.2.4 Pilot Production Phase

This phase of the life-cycle serves as a bridge between the prototype phase and the production phase. The purpose of the pilot production phase is to identify and correct manufacturing problems with the equipment before full-scale production begins. This is the first opportunity for the production equipment to be evaluated in an extended customer environment, and it is commonly called a beta-site evaluation. In fact, it is the first time that the equipment is exposed to a customer's processes.

Design and reliability engineers evaluate the actual level of equipment reliability. From this, they can determine what needs to be accomplished to meet requirements in a cost effective manner. Prior to the production and operation phase of the life cycle, reliability and design engineers continue design reviews to evaluate equipment reliability and make the appropriate recommendations. This is the last opportunity to make design changes and other improvements before full-scale production.

7.2.5 Production Phase

This phase of the life cycle represents the time when units are produced and sold. All major reliability problems are identified and corrected prior to the production phase. A formal program is in place for collecting and analyzing field service data, identifying root causes, and implementing corrective actions.

After proper review, decisions are made for resource allocation and for continuous improvement in the reliability process. The supplier and customer function as partners in these efforts.

7.2.6 Phase Out Phase

The equipment product line approaches the end of its useful life during this final phase of the life cycle. The end of useful equipment program life for the manufacturer can occur due to obsolescence or to no demand. To remain competitive, the manufacturer has to make plans for the next generation of equipment before phasing out the current-generation production.

Basically, each new generation of equipment will go through the same life cycle. The information gained during the six phases of the life cycle is retained so that it can be used to improve future generations of similar or new equipment.

7.3 Applying Reliability Metrics During Equipment Life Cycle

Figure 7.2 shows the six equipment life cycle phases in relation to the use of the three application categories of reliability metrics (described in Chapter 3). The figure shows whether they should be derived or if they should be used for a comparison purpose in each phase. This figure provides a guideline for proper use of the reliability metrics throughout the equipment life cycle phases.

Application of Reliability Metrics	Feasibility Phase	Design Phase	Prototype Phase	Pre-Prod. Phase	Production Phase	Phase Out Phase
Desired Value	Derived	Used for Comparison	Used for Comparison	Used for Comparison	Used for Comparison	Used for Comparison
Theoretical Value		Derived	Used for Comparison	Used for Comparison	Used for Comparison	Used for Comparison
Observed Value			Derived	Derived	Derived	Used for Comparison

Figure 7.2 Proper Uses of Reliability Metrics Throughout Equipment Life Cycle Phases

FOR EXAMPLE:

- The goals and requirements are derived during the feasibility phase. They are used to compare with either theoretical or observed values during the rest of the life cycle phases.
- The theoretical metrics are derived during the design phase. They are used to compare either with goals or with observed values during the rest of the life cycle phases.
- The goals are not derived during the production phase.

Chapter 8
The Reliability Improvement Process and Its Applications

In the last chapter, we described equipment and equipment program life cycle phases. In this chapter, we will study a generic reliability improvement process (RIP) that is embedded in every equipment program life cycle phase.

8.1 The Reliability Improvement Process

The reliability improvement process is a five-step iterative process, as shown in Figure 8.1. The five steps are:

1. Establish reliability goals and requirements for equipment
2. Apply reliability engineering and improvement activities, as needed
3. Conduct an evaluation of the equipment or equipment design
4. Compare the results of the evaluation to the goals and requirements and make a decision to move either to the next step or to the next phase.
5. Identify problems and root causes.

The process then returns to Step 2. Steps 2 through 5 are repeated until the goals and requirements are met.

These steps are described in greater detail in Reference 1. The following is a summary.

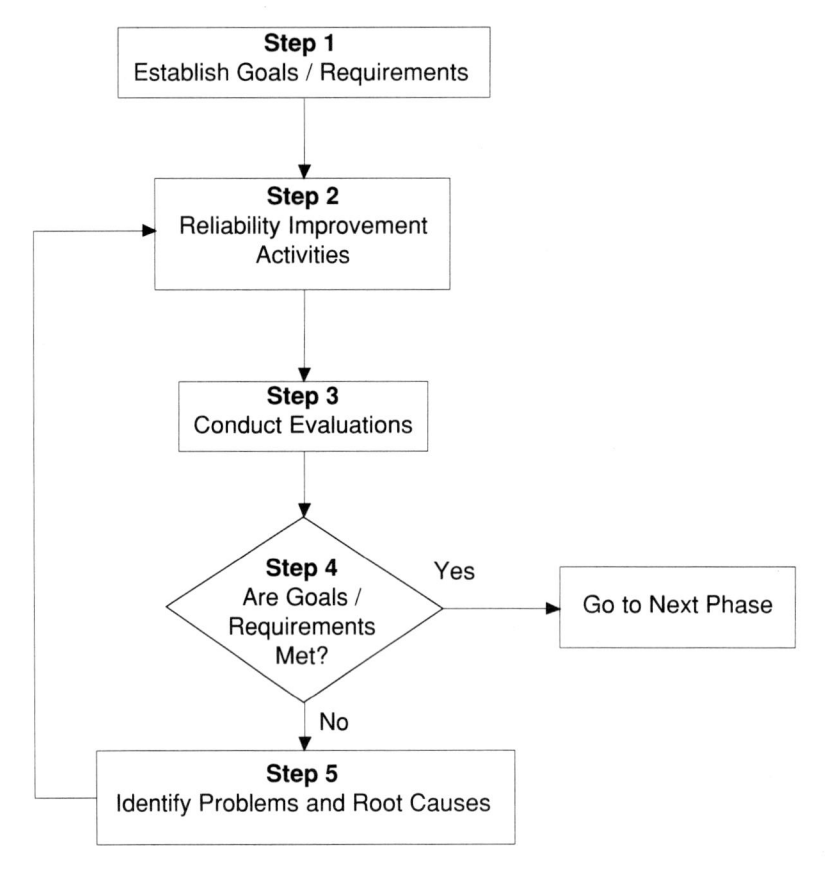

Figure 8.1 The Reliability Improvement Process

8.1.1 Establish Reliability Goals and Requirements

The first step in the reliability improvement process is to establish reliability goals and requirements. A distinction is made between goals and requirements, in that goals are more internally driven desires of an equipment manufacturer which may not be met. Requirements, on the other hand, are more specific and are customer driven. Requirements are usually included as deliverables in contractual agreements. They are expected to be met. Goals are the starting point, but are modified to satisfy customer requirements early in the equipment life cycle. The reliability goals and requirements must be attainable, supportable, acceptable, and measurable.

8.1.2 Reliability Engineering and Improvement Activities

Once goals and requirements are established, appropriate reliability improvement activities are applied to enhance the reliability of equipment. The activities depend upon the life cycle phase.

Basic practices that are applied to improve reliability in this step are:
- Designing in reliability
- Building in reliability
- Managing reliability growth
- Eliminating known causes of failure

8.1.3 Conduct an Evaluation

The next step in the reliability improvement process is to conduct an evaluation to assess the equipment reliability level. When equipment is in the early phases of its life cycle, reliability modeling is used to assess its reliability level. When equipment is in the pilot production or production and operation phases, test and field failure data analysis are used to assess its reliability level.

8.1.4 Determine Whether Goals and Requirements Are Being Met

Results of the evaluation process are compared to reliability goals and requirements. If goals and requirements are not met, the problems and root causes are identified, as described in the next section, and reliability improvement activities are initiated. If goals and requirements are met or exceeded, then approval is given to move to the next phase of the life cycle, where the reliability improvement process is again initiated.

8.1.5 Identify Problems and Root Causes

If reliability goals and requirements are not met, the reasons are identified and corrective actions are taken. Predictive modeling, test data on prototypes, or actual equipment field data are used to identify causes of failure and any potential reliability problems. The reliability improvement process then returns to Step 2, where reliability improvement and growth activities are initiated or upgrades and modifications to reliability goals and requirements are made.

Steps 2 through 5 are repeated until goals and requirements are met. The process requires several iterations of goal setting, evaluating, comparing, and improving. Approval is then given to move to the next phase of the life cycle, where the reliability improvement process is again applied.

8.2 Applying the Reliability Improvement Process

Optimal benefits from using the reliability improvement process occur when the process is applied to equipment during the concept and feasibility phase of the life cycle and then continuously thereafter. Benefits are also realized when the improvement process is applied to equipment that is in some advanced phase of its life cycle. Either way, it is important to address equipment reliability throughout the life cycle.

For example, reliability improvements may be necessary:

- Following the prototype phase because of design deficiencies or parts problems uncovered during prototype testing
- At the beginning of the pilot production phase because of reliability-related issues resulting from manufacturing a new equipment line
- During the production and operation phase because feedback from field personnel and customers indicates reliability problems caused by unanticipated failure mechanisms

8.3 Activities Associated with the Reliability Improvement Process

Activities associated with applying the reliability improvement process to the different phases of the equipment life cycle are listed in the Appendix A, Tables Al through A6. They are also described in detail in Reference 1. The activities remain basically the same from one phase of the life cycle to the next. The activities used may vary, however, depending on whether the improvement process has been continuously applied through the life cycle or whether it is being applied for the first time at an advanced phase.

For example, consider equipment in the prototype phase. If the reliability improvement process has been applied continuously to the equipment since the concept and feasibility phase, then the reliability goals and requirements already exist. Thus, the reliability goals and requirements activity consists primarily of updating the goals and requirements, and the primary focus is on prototype testing and corrective action activities. However, if the reliability improvement process is being applied to equipment for the first time during the prototype phase, then developing reliability goals and requirements would be a major focus, since they do not already exist.

8.4 Reasons to Use a Reliability Improvement Process

Knowledge of the equipment life cycle and the reliability improvement process in each phase is important because it provides a basis for understanding how and where reliability improvement activities should enter into the process of designing, producing, and operating the equipment. This cycle and process provide a framework for equipment suppliers to track reliability and to guide when and where to apply resources. The reliability improvement process provides a means for systematically improving reliability throughout the equipment life cycle. It drives design improvements, operations and maintenance procedure improvements, data collection and statistical analysis.

The basic structure of the reliability improvement process is being used by many semiconductor manufacturing equipment suppliers with significantly improved reliability over the life-cycle phases.

8.5 Reliability Plans

The equipment supplier should develop two-tier reliability plans based on the reliability improvement process, a general company-level plan covering all product lines, and a specific product-level plan, one for each equipment product line.

8.5.1 General Company Level Reliability Plan

This is an overall reliability plan tailored to the company. It takes into account the company's size, product lines, and available resources. The plan must address, at minimum, the following issues:

- The company's reliability policy
- A description of organizational activities (see Chapter 13)
- Identification of reliability champions
- Overall long-term strategy
- Acquisition and development of reliability skills within the company

8.5.2 Product Level Reliability Plans

Each equipment line should have a reliability plan based on the reliability improvement process activities summarized in Table 8.1. The plan should identify the product line's specific reliability activities, responsibilities, schedules, and resource requirements. This plan should be coordinated with the overall company-level reliability plan and the product program plan.

Table 8.1 List of Reliability Improvement Activities Throughout Equipment Life Cycle Phases

Concept & Feasibility Phase	Design Phase	Prototype Phase	Pilot Production Phase	Production Phase	Phase Out Phase
Goal Setting	Design-in Reliability	System Test	FRACAS	FRACAS	FRACAS
Apportionment	Modeling/Design Assessment	Reliability Level Assessment	System Test	PRAT	Transfer Reliability Knowledge
Reliability Plans	FMEA/FTA	Design Review	Life Test	Reliability Level Assessment	
Preliminary Modeling	Part Life Tests	FRACAS	Accelerated Test	Build Rel. Database	
	Design Review	Update CoO Calculations	Reliability Level Assessment	Update CoO Calculations	
	CoO Calculations		Design Review	Validate Modeling	

Note:
CoO = Cost of Ownership
FMEA = Failure Mode and Effects Analysis
FRACAS = Failure Reporting, Analysis and Corrective Action System
FTA = Fault Tolerance Analysis
PRAT = Product Reliability Acceptance Test

REFERENCE

1. SEMATECH, *Guidelines for Equipment Reliability*, Technology Transfer # 92031014A GEN, SEMATECH, Inc., Austin, TX, May 1992.

Chapter 9
Four Steps to Better Equipment Reliability

Not all equipment programs may go through all of the equipment life cycle phases described in Chapter 7. Some of the activities of the reliability improvement process may not be performed. However, if an equipment manufacturer follows the four basic steps to better equipment reliability, then there is a good chance that its equipment reliability will be higher than that of manufacturers who are not doing so. These steps are:

1. Know goals and requirements
2. Design in reliability
3. Build in reliability
4. Manage reliability growth

9.1 Know Goals and Requirements

If what is required is unknown, then it probably will not be achieved. Therefore, the first step to better reliability is to know the reliability goals and requirements, whether you are a manufacturer or a customer.

If you are a manufacturer (supplier) of equipment, you need to

- Understand the exact reliability requirements of your customer
- Be aware of the reliability level of your competitor's product
- Know what reliability level is required in the marketplace

Considering the above input, set the reliability goals for the equipment line at the beginning of each equipment program.

If you are a customer, then it is your responsibility to make sure that your equipment supplier knows your exact requirements.

The goals or requirements should include the following:

1. Reliability level and metric (e.g., MTBF = 700 hours)
2. Time factor, such as the age of the equipment at which it should attain the reliability level (e.g., 4 months after installation)
3. Operational conditions, such as:
   - Temperature and Humidity (e.g., Temperature range: 70–75°F, Humidity range: 40–50% RH)
   - Duty cycle (e.g., 12 hours/day)
   - Throughput rate (e.g., 15 wafers/hour)
   - Process to be used (e.g., high density plasma etch)
   - Operator skill level (e.g., grade 12 or equivalent)
   - Preventive Maintenance Policies to be followed (e.g., monthly PM policy described in the user's manual)
4. Shipping and installation limitations (e.g., to be shipped by air cushioned truck and installed by a special installation team)
5. Confidence level for the reliability metric (e.g., 80% confidence in the MTBF value)
6. Acceptable evidence for attaining the required reliability level (e.g., theoretical or calculated values; based on in-house test data, or based on field data)

Before you finalize the goals or requirements, make sure that every reliability goal or requirement is attainable.

9.1.1 Goal Allocation

Once the goals are known, the equipment manufacturer must break down the equipment and system-level goals into "bite-size" goals for subsystems, modules, and components. This makes it easy for subsystem, module, or component engineers to achieve their respective product goals.

The process of breaking down the equipment and system-level goal into the next levels of sub-goals, based on some logical justification, is called apportionment, budgeting, or allocation. This process is just like breaking down division-level budgets into department-level budgets. Many methods are available. Some widely used ones are:

Equal Allocation: Every lowest level component gets an equal share of the goal.

AGREE Allocation Method: Weight factors based on the complexity and criticality of the components are added in the calculations. See Reference 1 for details.

ARINC Allocation Method: Weight factors are based on the inherent failure rate. See Reference 1 for details.

9.2 Design in Reliability

This is the most important and elaborate step to achieve better equipment reliability. To design in reliability means to consider reliability improvement goals concurrently with other technical aspects at every activity of the design phase. Figure 9.1 depicts the process of designing in reliability. Six major blocks of the process are:

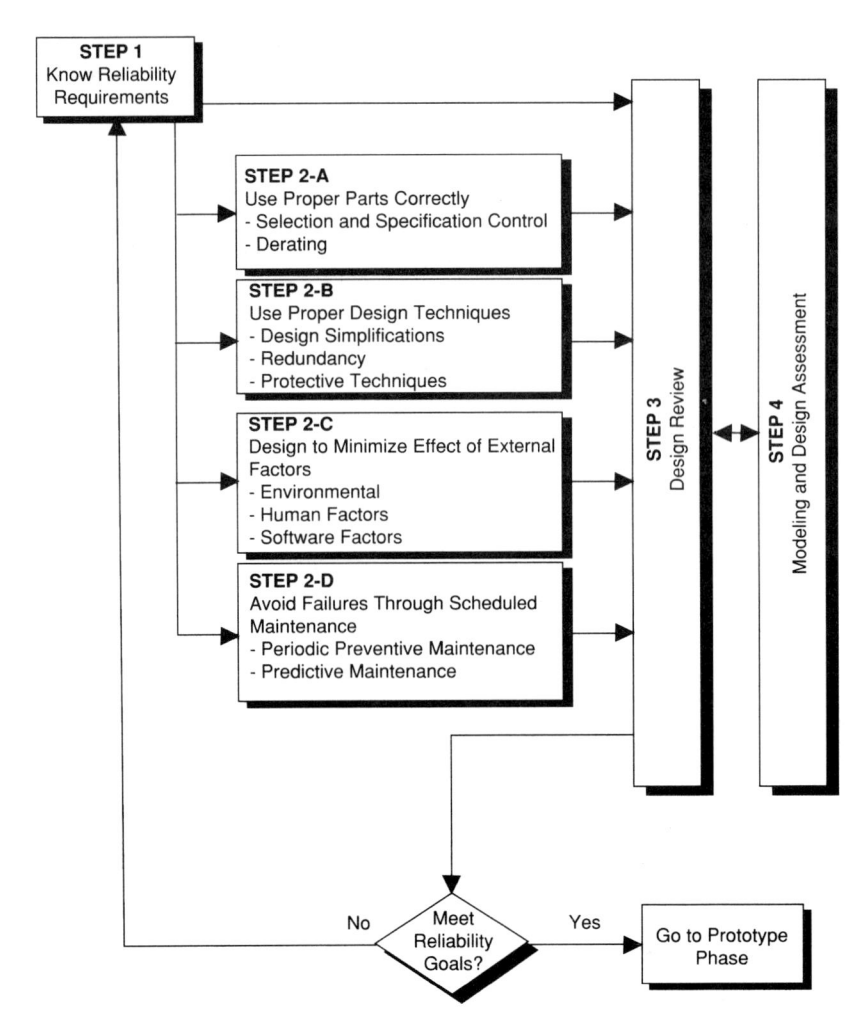

Figure 9.1 Process of Designing-in Reliability

1. Use the proper parts correctly
2. Use appropriate design techniques
3. Design to minimize the effect of external factors
4. Avoid failures through scheduled maintenance
5. Hold design reviews

6. Assess the reliability of the design using the modeling techniques

Reference 2 contains a detailed description of each block. The following is a brief summary.

9.2.1 Use Proper Parts Correctly

Use of proper part correctly is the most crucial block of the process to design in reliability. It consists of the following activities:

Part Selection: Before selecting any part and its supplier, determine the part type needed to perform the required functions and the environment in which it is expected to operate. The general rule for part selection is that, whenever possible, the designer should strive to use proven parts in the design and select a supplier who has proven historically to meet or exceed the part reliability requirements.

See Chapter 12 for details on how to purchase reliable parts.

Part Specification: For each reliability sensitive part, the procurement specification should include:

- Details of the intended application(s)
- Reliability requirements of the intended application(s)
- Part screening procedure
- Part qualification procedure

Part Derating: Once the part is selected, perform an analysis to compare the expected stress level for the intended applications with those of the part's rated (capacity) stress level. A technique known as *derating* is used to improve the design reliability. In this technique, a part is selected so that it will operate at less severe stress than the stress level that it is rated capable of operating. For example, if the expected power level is 10 watts for a device, select the parts that are rated for significantly higher than 10 watts power. Use the appropriate derating factors for various electronics components given in Table 9.1.

Table 9.1 Derating Factors for Electronic Parts

Component	Stress Category	Derating Factor
Capacitors, General	Voltage	0.5
Ceramic Capacitors	Voltage	0.5 at < 85°C 0.3 at < 125°C
Supermetallized, Plastic Film, Any Tantalum Capacitors	Voltage Temperature	0.5 at < 85°C less than 85°C
Glass Dielectric, Fixed Mica Capacitors	Voltage Temperature	0.5 at < 85°C less than 85°C
Connectors	Current Voltage Temperature	0.5 0.5 less than 125°C
Quartz Crystals	Power	0.25
Diodes	Voltage Temperature	0.75 0.5
EMI & RFI Filters	Voltage Current	0.5 0.75
Fuses	Current	0.7 at < 25°C 0.5 at < 25°C
Integrated Circuits (All Kinds)	Voltage Current Power	0.7 0.8 0.75
Resistors (All Kinds)	Voltage Power	0.8 0.5
Thermistors	Power	0.8
Relays & Switches	Current	0.75 for resistive load 0.4 for Inductive load 0.2 for motors 0.1 for filament
Transistors	Breakdown Voltage Junction Temp	0.75 less than 105°C
Wires And Cables	Current	0.6

Note: Above factors were derived from Reference 3

9.2.2 Use Proper Design Techniques

Use of proper design techniques is another crucial block of the design-in reliability process. It consists of the following activities:

Design Simplification: Anything that can be done to reduce the complexity of the design will, as a general rule, improve reliability. If a part is not required, eliminate it from the design. Wherever possible, reduce the number of parts by combining functions.

Redundancy: This is one of the most popular methods in design to achieve the needed level of reliability. Redundancy is the provision of more than one part for accomplishing a given function so that all parts must fail before causing a system failure. Redundancy, therefore, permits a system to operate even though some parts have failed, thus increasing system reliability.

For example, if we have a simple system consisting of two identical redundant (parallel) parts, the system MTBF will be 1.5 times that of the individual part MTBF (see Chapter 4).

Protective Technique: This technique includes a means in the design to prevent a failed part or malfunction from causing further damage to other parts. The following are some of the popular protective techniques used in equipment designs:

- Fuses or circuit breakers to sense excessive current drain and to cut off power to prevent further damage
- Thermostats to sense over-temperature conditions and shut down the part or system operation until the temperature returns to normal
- Mechanical stops to prevent mechanical parts from traveling beyond their limits
- Pressure regulators and accumulators to prevents pressure surges
- Interlock to prevent inadvertent operations

9.2.3 Minimize Effect of External Factors

The operating environment is neither forgiving nor understanding. It methodically surrounds and affects every part of a system. If a part cannot sustain the effects of an expected environment, then reliability suffers. First, the equipment manufacturer must understand the operating environment and its potential effects. Then, it must select designs and materials that counteract these effects or provide methods to alter and control environmental conditions within acceptable limits.

Equipment design engineers must consider the following external factors affecting reliability:

- Heat generation causing high temperatures
- Shock and vibration
- Moisture
- High vacuum
- Explosion
- Electromagnetic compatibility
- Human use
- Software design

9.2.4 Avoid Failures Through Scheduled Maintenance

One way to improve reliability is to minimize the number of failures that occur during operation. This can be achieved in two ways:

1. Select parts that fail less frequently
2. Replace a part before its expected failure time

The latter method is known as *scheduled maintenance* (SM). This technique is used when it is not feasible to find a part that fails less frequently. If such a situation is properly comprehended during the design phase, it can be avoided through one of the following SM techniques.

Periodic Preventive Maintenance: This is a fixed-period-driven maintenance procedure in which the parts that are partially worn out, aged, out-of-adjustment, or contaminated are replaced, adjusted, or cleaned before they are expected to stop functioning. This way, the system failures are forestalled during the system operations, thus reducing the average failure rate.

Predictive Maintenance: This is a condition-driven scheduled preventive maintenance program. Instead of relying on fixed-period-of-life units to schedule maintenance activities, predictive maintenance uses direct monitoring of appropriate indicators to determine the proper time to perform the required maintenance activities.

9.2.5 Design Review

Design reviews are an essential element of the design-in reliability process. The main purposes of a design review are to ensure that:

- Customer requirements are satisfied
- The design has been studied to identify possible problems
- The alternatives have been considered before selecting a design
- All necessary improvements are based on cost trade-off studies

Conduct design reviews on a regular basis from the design feasibility study through the pilot production phase. An effective design review team should have representation from each functional area involved in developing the equipment. Table 9.2 contains the make-up of an effective design review team and the responsibility of each member. To facilitate the design review process, use the appropriate system-level or part-level checklists from Table 9.3 or 9.4.

Table 9.2 Design Review Team Members and Their Responsibilities

Member	Responsibilities
Team Leader*	Coordinate and conduct meetings; issue minutes of the meeting and interim and final reports
Design Engineers (Mechanical, Electrical, and Industrial)	Prepare and present design and substantiate decisions with data from tests or calculations
Reliability Engineer	Evaluate design for optimum reliability, consistent with goals
Manufacturing Engineer	Ensure that the design is producible at affordable cost and schedule
Service (Field) Engineer Maintainability Engineer	Ensure that installation, maintenance and operation considerations are included in the design
Procurement Representative	Assure that acceptable parts and materials are available to meet cost and delivery schedule
Quality Engineer	Ensure that the functions of inspection, control, and test can be efficiently carried out
Material Specialist	Ensure that the material selected will perform as required
Process Engineer	Ensure that the hardware and software will be capable of meeting process requirements
Software Engineer	Ensure that the hardware and software will be compatible and will optimize reliability
Safety Engineer	Ensure that safety issues are resolved
Tool Engineer	Evaluate design in terms of the tooling costs required to satisfy tolerance and functional requirements
Packaging and Shipping Specialist	Assure that the product is capable of being handled without damage
Marketing Representative	Assure that the customers' requirements are met
Toughest Customer (optional)	Include ultimate users' concerns

* Preferred Team Leader is manager of Design Engineering or Reliability Engineering

Table 9.3 System Level Checklist for Design Review Team

#	Item
1.	Have specific design criteria for system level reliability been established?
2.	Have acceptance, qualifications, sampling, and reliability assurance testing been established?
3.	Have specific maintenance criteria for maintainability been established?
4.	Is the design simple? Does it have minimum number of parts?
5.	Is it designed as a unified overall system rather than an accumulation of parts, etc?
6.	Are standard high-reliability parts being used?
7.	Are high failure-rate parts identified? Has a PM policy been developed for these parts?
8.	Have limited-life parts been identified and PM requirements specified?
9.	Have critical parts that require special procurement, testing, or handling been identified?
10.	Has redundancy been provided where needed to meet specific reliability goals?
11	Are there adequate indicators to verify critical functions?
12.	Is there a provision for improvements to eliminate design inadequacies observed in tests?
13.	Have adjustments been minimized and made accessible?
14.	Have stability requirements for all parts associated with each adjustment been established?
15.	Is there a concentrated effort to make the development model as near to the production model as possible?
16.	Are there adequate safety precautions taken to satisfy the safety requirement?
17.	Is the design of software control and user interface correct and appropriate?

Table 9.4 Part Level Checklist for Design Review Team

#	Item
1.	Is the part compatible with system to be designed?
2.	Are the following requirements established for each reliability sensitive part? - Reliability level (failure rate, MTBF, etc.) - Reliability test - MTTR - Stress levels - Process capability and accuracy requirements for the process stabilizing parts
3.	Is the part a standard high reliability part? If not, can the selected part meet the reliability requirements?
4.	Is the reliability (MTBF) of the part based on actual application of the part?
5.	Has the shelf life of the part been determined?
6.	Have levels of different stress factors been determined?
7.	Have derating factors been used in application of the part?
8.	Have safety factors and safety margins been used in application of the part?
9.	Are the best available methods for reducing the adverse effect of operational environments on the part being utilized?
10.	Have normal modes of failure and the magnitude of each mode for the part been identified?
11.	Have the PM requirements for the part been identified?
12.	Is the part properly integrated and installed in the system?

9.2.6 Reliability Assessment of the Design

Once the equipment design starts taking shape, it must be assessed to determine its reliability level and the system level effect of failure rate of each part. Modeling techniques determine analytical value of the reliability level. Failure Mode and Effects Analysis (FMEA) technique determines and alleviates the system-level effect of part failures. There are many commercially available software packages that perform modeling and FMEA.

9.3 Build-in Reliability

Building-in reliability is a process that assures that parts, subsystems, and modules are made and assembled according to engineering drawings and specifications without degrading the designed reliability or introducing new failure modes. Important steps of this process are:

Assembly Instructions: Prepare detailed instructions for each assembly step. These instructions should include proper parts, materials, step-by-step assembly procedures, tools, limitations, inspection procedures, etc.

Training: To minimize the assembly errors, it is essential that every assembly operator is trained in basic assembly methods and in all the assembly operations assigned to him or her.

Burned-in: All parts and the system itself should be properly and adequately burned-in, debugged, or stress-screened before shipment.

Product Reliability Acceptance Test (PRAT): Conduct a PRAT on randomly selected units before shipping to assure the reliability level of the product line as it is being shipped.

Packaging and Shipping: The equipment must be packed properly for the intended shipping mode. Select the shipping mode so shipping does not impart any undue stress on the equipment.

9.4 Manage Reliability Growth

Effectively managing reliability growth opportunities is a continuous improvement process. Equipment manufacturers learn from actual in-house tests or field experience. All observed problems are documented and brought back to a central body for further analysis and disposition. If required, corrective actions are developed and implemented.

A popular system named "Failure Reporting Analysis and Corrective Action System" (FRACAS) is used to manage this process. As shown in Figure 9.2, FRACAS is a closed-loop feedback communication channel in which the customer works with the equipment supplier to collect, record, and analyze equipment failures, both hardware and software. The customer captures predetermined types of data about all problems observed with a particular equipment line and submits the data to that supplier. A Failure Review Board (FRB) at the supplier's site analyzes the failures. The resulting analysis identifies corrective actions that should be developed, verified, and implemented to prevent failures from recurring.

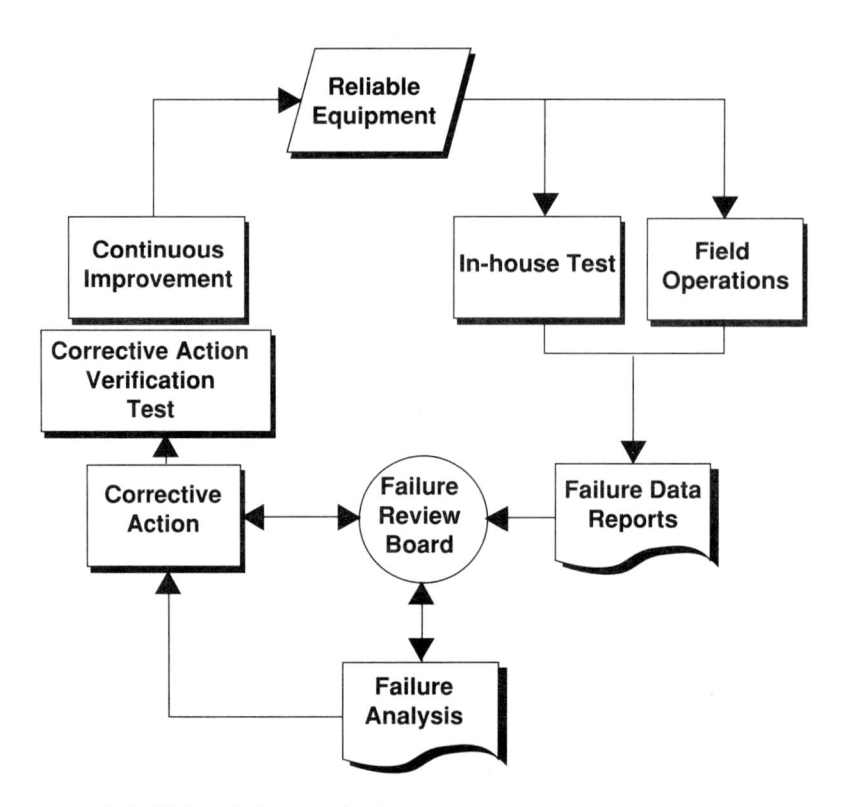

Figure 9.2 FRACAS Process Flow

Now let us look at three key elements of FRACAS in more detail.

Failure Data Reporting: All the failures, observed either during in-house test or at customer's site, must be recorded so all relevant and necessary data is captured in a systematic manner. A simple, easy-to-use form that is tailored to the respective equipment line should be used to record and report failure data. (Figure 9.3 depicts a typical failure reporting form). If the data volume justifies the cost of administering FRACAS, the data form can be computerized to communicate failure data. Some customers even use the Internet and electronic mail to report failure data.

SERVICE REPORT NO.

Customer	Customer System ID	Module	Received By:
Reported By	Equipment Serial No.	Module Serial No.	Ref. Report NO.
Phone No.	Contact	Date	Time

Service Category:
☐ Install ☐ Courtesy ☐ Warranty ☐ PM ☐ Development
☐ Training ☐ Telephone Fix ☐ Update ☐ UM ☐ Others

Problem/Symptoms/Reasons For Service Call

Repair & Maintenance Action(s)

Problem Cause

Part Time	Part Number	Replaced	Cleaned	Comments

Status of Equipment Upon Leaving	Start Date	Time
	Complete Date	Time
Customer Remarks	Service Engineer's Signature	
	Customer's Signature	

Figure 9.3 *A Typical Failure Report Form*

Chapter 9

Failure Review Board: The FRB is a multifunctional self-managed team that reviews, facilitates, and administers failure analysis. It also participates in assigning, developing, verifying, and implementing the resulting corrective actions. To do this job effectively, all the functional departments involved in the product line must participate on the FRB. Also, FRB members must be empowered to assume responsibility, investigate failure cause, and develop corrective actions.

Corrective Action: Any systematic action taken to eliminate or reduce the frequency of equipment failure (hardware or software) is a corrective action. Such actions may include parts design or materials changes, part supplier changes, assembly procedure changes, maintenance procedure changes, operational changes, training changes, or software changes.

When the cause of a failure has been determined, a corrective action plan is developed, documented, and implemented. The plan should identify the following three W's:

- What actions are to be taken?
- Who is responsible for each action?
- When will each action be completed?

All the corrective action plans and their verification and implementation should be reviewed by the FRB on a regular basis. The FRB should also maintain a log of the corrective action status including open and closed corrective actions.

REFERENCES

1. ARNIC Research Corp., *Reliability Engineering,* Prentice Hall, Englewood Cliffs, NJ, 1964.

2. SEMATECH, *Design Practices For Higher Equipment Reliability - Guidebook*, Technology Transfer #93041608A-GEN, SEMATECH, Inc., Austin, TX, 1993.

3. MIL-STD-1574A, *Electronic Parts, Material, and Processes for Space Launch Vehicles*, USAF, 1987.

Chapter 10
Three Reliability Growth Mechanisms

Reliability level of an equipment improves in the following three ways: (1) with age during the early life of equipment use cycle, (2) with the maturity of an equipment program throughout equipment program life cycle phases, and (3) from one equipment generation to the next generation. This means an aged equipment has better reliability than that of a new one, and an equipment manufactured later in the equipment program production phase is more reliable than that manufactured earlier. Also, the recent generation of equipment is more reliable than the earlier generation. This phenomenon of reliability growth over time has long been recognized and has been studied by many experts. They have developed many empirical formulas to model the growth mathematically (see Reference 1 and 2). This chapter describes the following three growth mechanisms and presents the most widely used mathematical model for each mechanism.

1. Reliability growth mechanism during the early life of an equipment
2. Reliability growth mechanism throughout the program life cycle phases of the equipment program
3. Reliability growth mechanism from one generation to the next generation

See Chapter 7 to understand the difference between equipment and an equipment program. Figure 10.1 depicts hierarchy of the three growth mechanisms.

Let us examine the three growth mechanisms in detail.

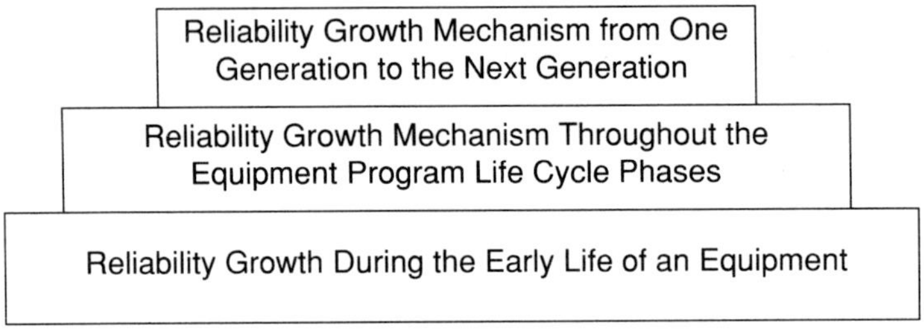

```
┌─────────────────────────────────────────┐
│  Reliability Growth Mechanism from One  │
│    Generation to the Next Generation    │
└─────────────────────────────────────────┘
┌───────────────────────────────────────────────┐
│  Reliability Growth Mechanism Throughout the  │
│     Equipment Program Life Cycle Phases       │
└───────────────────────────────────────────────┘
┌─────────────────────────────────────────────────────┐
│ Reliability Growth During the Early Life of an Equipment │
└─────────────────────────────────────────────────────┘
```

Figure 10.1 Hierarchy of the Reliability Growth Mechanisms

10.1 Reliability Growth Mechanism During the Early Life of Equipment

As shown in Figure 2.8, reliability improves (i.e., failure rate decreases) as an equipment gets older and then stabilizes at a constant rate. This growth period, as described in Chapter 7, is known as infant mortality period. Reliability growth during this period occurs by finding and removing manufacturing and workmanship defects.

Reference 3 has shown that the time to failure (in this period) follows a well known Weibull distribution with a shape parameter, β, of less than 1. The failure rate is given by

$$\lambda(t) = \frac{\beta}{\eta} = \left\{\frac{t}{\eta}\right\}^{(\beta-1)} \tag{10.1}$$

WHERE:

$\lambda(t)$ = Failure rate at age t during the early life period
β = Shape parameter, $0 < \beta < 1$, also known as reliability growth rate constant
η = Scale parameter, $\eta > 1$, also known as 36.8 percentile point

Use above model to determine realized reliability growth for units of equipment and to predict $\lambda(t)$ at time t for similar equipment. To determine the realized growth, collect failure rate $\lambda(t)$ at a given age t data for a sample of units of equipment and fit them to equation (10.1) to determine the least square fit values of growth parameters β and η (See Reference 2 for more details). For example,

t	λ(t)
50	1.740×10^{-3}
100	1.318×10^{-3}
200	1.000×10^{-3}
400	0.757×10^{-3}
1000	0.528×10^{-3}

The least square fit value for $\beta = 0.6$ and $\eta = 1,250$.

Once the growth parameters are known, use equation (10.1) to calculate predicted values of $\lambda(t)$ at any time t for the similar equipment (i.e., having similar configuration) during the early life period.

For example, if $\beta = 0.6$ and $\eta = 1,250$ then $\lambda(600) = 0.644 \times 10^{-3}$ failures per hour

Note that growth parameters change with the equipment program maturity as explained in the next section.

10.2 Reliability Growth Mechanism Throughout the Equipment Program Life Cycle Phases

Second reliability growth mechanism originates from all the continuous reliability improvement activities described in Chapters 8 and 9. This growth mechanism tells that an equipment manufactured later in the equipment (program) production phase is more reliable than that manufactured earlier. This growth occurs by finding and removing design defects, misapplied parts, manufacturing errors, software errors, and service procedure and training deficiencies. Figure 10.2 shows the difference between the first two reliability growth mechanisms.

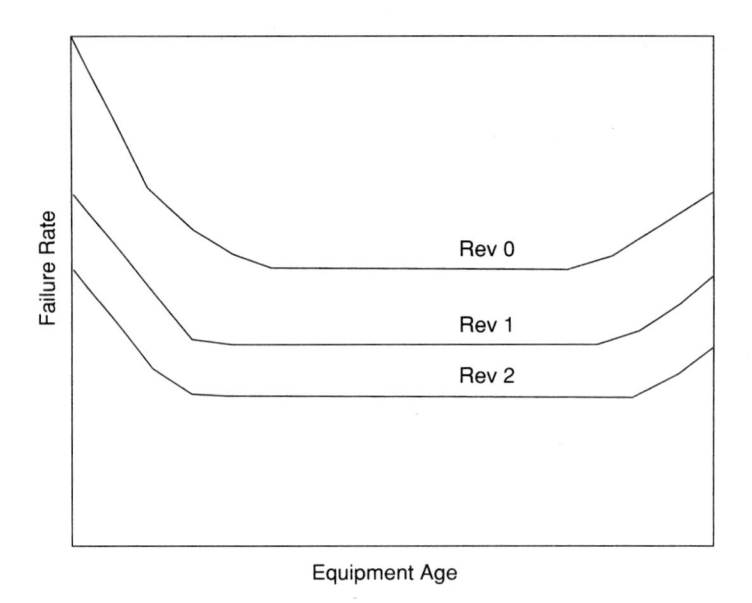

Figure 10.2 Reliability Growth Mechanisms

As shown in Reference 1 and 2, this growth mechanism is extensively studied and experts have developed several mathematical models to describe it. One of the models, more appropriate for an equipment program, is known as *Duane MTBF Growth Model*. This model is given as

$$\text{MTBF}(T_a) = \text{MTBF}(1) \times (T_a)^\alpha \qquad (10.2)$$

WHERE:

T_a = Accumulated hours of test and/or field experience
MTBF(T_a) = Cumulative MTBF after T_a accumulated hours of test and/or field experience
MTBF(1) = MTBF at T_a =1 or at the beginning of the test, or the earliest time at which the first MTBF can be determined
α = Reliability growth rate constant, $0 < \alpha < 1$

A closer look at equation (10.2) reveals that value of α drives the MTBF growth. Therefore, proper selection of value of α is very important. The value depends upon equipment type, quality of parts, and aggressiveness of the reliability improvement efforts. One way to determine α is to collect cumulative MTBF and corresponding T_a data and fit them to model equation (10.2) to determine the least square fit value of MTBF(1) and α. For example,

MTBF(200) = 100 hr
MTBF(400) = 133 hr
MTBF(600) = 150 hr, and
MTBF(3,000) = 273 hr

give the least square fit value of MTBF(1) = 14 and α = 0.376.

In the absence of such data, use Table 10.1 as a guideline to select appropriate value of α.

Table 10.1 Recommended Values of Reliability Growth Constant α

Types of Reliability Improvement Program	α
No formal reliability improvement program	0.20
Reactive informal reliability improvement program	0.45
Proactive formal reliability improvement program with failure reporting, analysis, and corrective action system in place	0.65

Use the least square fit or tabulated values in equation (10.2) to calculate predicted value of MTBF(T_a) for the equipment manufactured after T_a hours of accumulated test and/or field experience.

For example, if MTBF(1) is 14 hours, α = 0.376, and T_a = 5,000 hours, then the expected cumulative MTBF after 5,000 hours of accumulated test is 344 hours.

Chapter 10

The reliability growth curve is a powerful tool for managing a continuous improvement process and predicting reliability level for future configurations.

10.3 Reliability Growth Mechanism From One Generation to the Next Generation

Our experience shows that every time we create a new generation of equipment, the new generation is more reliable than the older one. This reliability growth mechanism originates from all the continuous reliability improvement activities learning and transferring reliability improvement knowledge to the next generation, and technological advances. So far, no systematic studies have been conducted to formulate this growth mechanism. As a result, there are no mathematical models that quantify this growth mechanism. In the absence of such studies, we can use models given in equations (10.1) and (10.2), with higher values of growth parameters α and β, to predict reliability of an equipment manufactured during an equipment program in the next generation of equipment.

REFERENCES

1. R. E. Schafer, R. B. Sallee, and J. D. Torrez, *Reliability Growth Study*, Hughes Aircraft Company, Fullerton, CA, 1983.

2. Dimitri Kececioglu, *Reliability Engineering Handbook*, Volume 2, PTR Prentice Hall, Englewood Cliffs, NJ, 1991.

3. Dimitri Kececioglu, *Reliability Engineering Handbook*, Volume 1, PTR Prentice Hall, Englewood Cliffs, NJ, 1991.

Chapter 11
Reliability Testing

No matter how many or how extensive the analyses we perform to calculate the reliability level of equipment[1], it is almost impossible to calculate the effect of all the factors that affect the reliability level. Even after engaging fancy reliability modeling software programs to do the reliability level calculations, we cannot theoretically derive the exact reliability level that will be observed in a reliability test or when equipment is installed at the customer's site. This lack of confidence in our theoretical efforts necessitates performing reliability tests to find the actual reliability level of the tested equipment configuration.

Reliability testing is a very important activity of a reliability improvement program. The tests provide the proof for all the theoretical calculations and promised performance indices. Information generated during the reliability test is vital to design engineers for initial designs and subsequent redesigns or refinements, and to manufacturing engineers for fine tuning the manufacturing process. The reliability tests also provide vital information to program managers showing technical progress and problems of an equipment line.

Reliability tests can be performed at any level of integration, i.e., at the component level, part level, module level, subsystem level, or system level. Not only that, they can be performed during any equipment program life cycle phase.

Among the numerous reasons to conduct reliability tests are:

---

1 The text of this chapter refers to equipment only. However, the reliability tests described in this chapter apply equally well to parts, subsystems, and modules.

- To determine the reliability level under the expected use conditions
- To qualify that the equipment line meets or exceeds the required reliability level
- To ensure that the desired level is maintained throughout the equipment life cycle phases.
- To improve reliability by identifying and removing root failure causes

11.1 Types of Reliability Tests

Since reliability testing is included in all the equipment life cycle phases, and it is conducted for numerous reasons, it follows that the testing includes many types of tests. The following reliability tests are commonly seen during a typical equipment program.

1. Burn-in test
2. Environmental stress screening (ESS) test
3. Reliability development/growth test
4. Reliability qualification test
5. Product reliability acceptance test (PRAT)
6. Accelerated test

Now let us explore each type in a little more detail.

11.1.1 Burn-in Test

This test is conducted to screen out parts that fail during the early life period (see Figure 2.8). It is performed at part, subsystem, or system level. Most failures observed during this test are due to manufacturing workmanship errors, poor quality parts, and shipping damage. The system level burn-in tests are also known as debug tests.

11.1.2 Environmental Stress Screening Test

As the title indicates, the ESS tests are conducted in an operating environment that is more harsh (stressed) than the normal environment for expected use. The main purpose of the test is to weed out parts that otherwise would not fail under normal operating environment. This test increases confidence that all received parts are of good quality and they will last longer (i.e., have better reliability).

11.1.3 Reliability Development/Growth Test

The reliability development/growth tests are conducted to ensure that a desired reliability level is achieved during a given equipment program life cycle phase and it is improving (growing) as the program moves further in the life cycle phases. Most of these tests are run at the system level.

11.1.4 Reliability Qualification Test

This test is conducted to qualify that the part or equipment meets or exceeds the reliability level. This is a pass-fail test. If the demonstrated reliability level is equal or better than the required level, the equipment (or its program) is considered as meeting the requirement, passing the test or qualifying the situation.

11.1.5 Product Reliability Acceptance Test

This test is very similar to the Reliability Qualification Test, except it is conducted on equipment randomly selected from those that are ready to ship to customers.

11.1.6 Accelerated Test

Reliability development or qualification tests are ordinarily too drawn out to provide the needed information quickly enough to make decisions or to permit changes. To circumvent this, many equipment manufacturers employ a testing technique called "Accelerated Testing." In this technique, operational stresses are increased so that the expected failure will arrive in a shorter time than it takes under normal operational stresses. This way, we compress the calendar time, that is, we accelerate the tests. Once the equipment life (reliability level) is determined under the higher operational stresses, the observed life is interpreted for normal operating environment. The most commonly used acceleration techniques are listed below.

- Enlarge the sample size (e.g., test 8 units instead of 3)
- Increase usage rate (e.g., run test at 100 cycles per hour instead of normal rate of 40)

- Increase operational stresses (e.g., run test with 100 lb. load instead of 40 lb.)
- Increase environmental stresses (e.g., run at 120° C. instead of at 30°C)
- Combine any of the above techniques

11.2 Generic Steps for Reliability Tests

Three overall steps of any typical reliability test are:

1. Test plan development
2. Test conducting
3. Test data analysis and reporting

11.2.1 Test Plan Development

A well thought-out reliability test plan includes the following:

- Test objectives
- Hardware and software to be used
- Operational stresses and environment
- Resources required (including consumable)
- Sample size and test length
- Test procedure
- Data to be acquired
- Data form to be used
- Data analysis techniques
- Data reporting and reviewing procedures
- Pass-fail criteria, if required
- Expected outcome for each test
- Types of test reports
- Schedule of key test activities

The reliability test plan should be formally documented and approved by high level managers.

11.2.2 Test Conducting

During this step, the reliability test is conducted according to the test plan. All deviations from the formal test plan should be recorded and approved. A formal log of test events is kept to record key test parameters associated with each event.

11.2.3 Test Data Analysis and Reporting

All the data collected during the test are appropriately analyzed, and conclusions are made. Test data, results, and conclusions should be reviewed by the FRB and other interested groups. To close a test project, a formal test report must be issued containing test objectives, test procedures, findings, conclusions, and recommendations.

11.3 Reliability Tests Throughout the Equipment Program Life Cycle Phases

As shown in Chapter 8, reliability tests are scattered throughout the equipment program life cycle phases. They play a very important part in the reliability improvement process. Table 11.1 lists the appropriate tests for each phase.

Table 11.1 Reliability Tests Throughout the Equipment Life Cycle Phases

Life Cycle Phase	Reliability Test
Concept and Feasibility	No Formal Reliability Test
Design	Part-Level Reliability Qualifications Reliability Development Accelerated Test
Prototype	Part-Level Reliability Qualifications Reliability Qualifications Reliability Growth Accelerated Test
Pilot Production	Burn-in Environmental Stress Screening System-Level Reliability Qualifications Accelerated Test Reliability Growth
Production	Burn-in Environmental Stress Screening Reliability Qualifications Product Reliability Acceptance Test Accelerated Test Reliability Growth
Phase Out	None Recommended

11.4 Test Length

The test length depends upon the desired confidence in the test results and the expected level of reliability (MTBF). Tests need to run long enough to increase confidence in the test results. However, we never have enough resources or time to test for an extended period. Therefore, statisticians have developed a method to determine the minimum test length needed to make correct decisions with the required confidence in that decisions. There are many test length tables available to fit any test circumstances. See References 1 and 2 for such tables. For repairable equipment, minimum test lengths are determined to obtain the certain minimum MTBF level (target MTBF) with certain confidence, by

Minimum Test Length with P% Confidence = (Target MTBF) × ω   (11.1)

WHERE:

P% = Desired confidence level
Target MTBF = MTBF to be proved or expected
ω = Appropriate multiplier for P% from Table 11.2

For example, if we need to prove target MTBF of 100 hours, with 80% confidence in the decision, the minimum test length is calculated as follows.

Target MTBF = 100 hours
ω = 1.61 from Table 11.2 for 80% Confidence Level

Table 11.2 Minimum Test Length Multiplier ω

	\multicolumn{7}{c}{Confidence P}
	10%	20%	50%	75%	80%	90%	95%
Multiplier ω	0.11	0.22	0.69	1.38	1.61	2.30	2.99

These give a minimum test length = 100 × 1.61 = 161 hours.

11.5 Test Data Analysis

Reliability textbooks contain many formulas for determining the observed reliability level and the associated confidence limits. For our repairable equipment, the following simplified method can determine lower confidence limits for MTBF with the desired confidence level.

P% Lower Confidence Limit for the MTBF = (Observed MTBF) × K

(11.2)

WHERE:

- P% = Desired confidence level
- Observed MTBF = MTBF calculated based on the test length and number of failures observed
- K = Appropriate multiplier, from Table 11.3, for the number of failures observed during the test and the desired confidence level

Table 11.3 Multiplier K for the Confidence Limit Calculations

Number of Failures	\multicolumn{6}{c}{Confidence P}
	70%	80%	85%	90%	95%	97.5%
1	0.710	0.620	0.530	0.434	0.333	0.270
2	0.734	0.667	0.600	0.515	0.422	0.360
3	0.766	0.698	0.630	0.565	0.476	0.420
4	0.786	0.724	0.662	0.598	0.515	0.455
5	0.812	0.746	0.680	0.625	0.546	0.480
10	0.848	0.800	0.752	0.704	0.637	0.585
15	0.872	0.826	0.790	0.746	0.685	0.640
20	0.884	0.847	0.810	0.768	0.719	0.675
30	0.900	0.870	0.840	0.806	0.756	0.720

Note: The K factors in this table apply to failure truncated tests only. For time truncated tests see Reference 1.

For example:

If three failures were observed during a 10,000 hour reliability test, then

Observed MTBF = 10,000/3 = 3,333 hours (11.3)

K Factor for 3 failures and 80% confidence level is 0.698 from Table 11.3, therefore, using equation (11.2)

80% Lower Confidence Limit = 3,333 x 0.698 = 2326 hours. (11.4)

REFERENCES

1. W. Grant Ireson, *Reliability Handbook*, McGraw-Hill, New York, NY, 1966.

2. Robert E. Odeh and Martin Fox, *Sample Size Choice*, Marcel Dekker, New York, NY, 1975.

Chapter 12
How to Buy Reliable Equipment and Parts

Buying was once a simpler activity—just negotiate price and write a purchase order. Times have changed this activity into an essential ingredient of the reliability improvement process. How you buy equipment for your use makes a big impact on the reliability level of the equipment. Similarly, how you buy parts makes a big impact on the reliability level of the equipment you manufacture and sell.

In this chapter, we will explore ways to use buying activities to improve the reliability level of the purchased or manufactured equipment.

As mentioned earlier, everyone working on a product line contributes to and is responsible for achieving reliability goals. Parts or equipment buyers are no exception to this. They play a crucial role in achieving the reliability goals.

To take full advantage of the buying activities, a company should implement a formal buying process similar to that shown in Figure 12.1. This process effectively integrates reliability improvement activities with buying activities.

Let us examine each process step in detail.

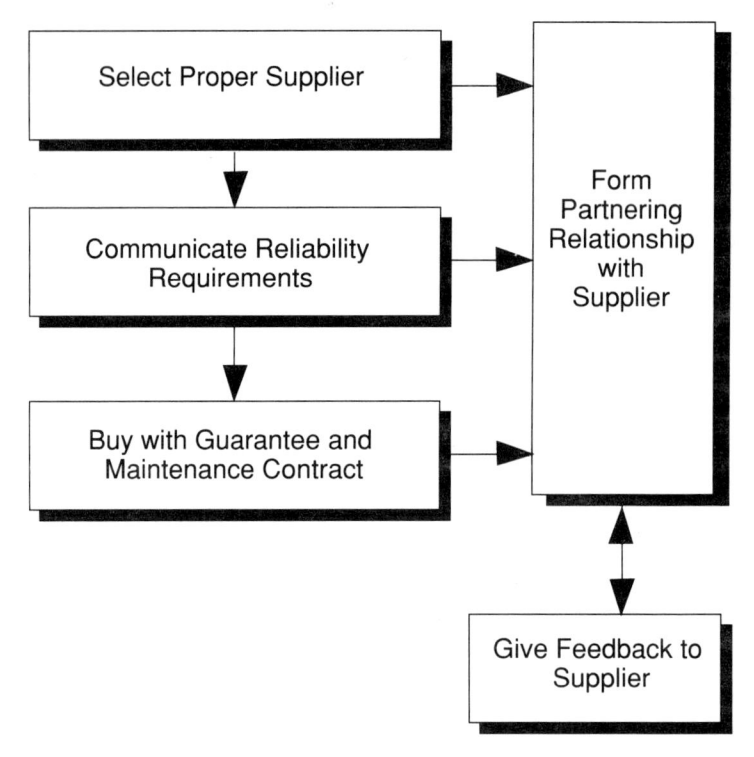

Figure 12.1 Buying Process with the Reliability Improvement Activities

12.1 Select Proper Supplier

Supplier selection is the first step of acquiring anything. If you wish to buy reliable equipment or parts, the general rule for supplier selection is that, whenever possible, you need to select a supplier who is known and has a reputation for supplying reliable products. If the supplier being considered has supplied your company in the past, review the history of quality and reliability of that product. Select a supplier for additional purchase only if you are satisfied with the past performance.

If you are considering a supplier with whom you have never worked, make sure that the supplier is reputable, with an effective quality system in place to control the quality and reliability of its product. One way to find out whether a sound quality system is in place is to look for the following indicators:

- Supplier is an ISO 9001 compliant
- Supplier recently received Malcolm Baldrige National Quality Award
- Supplier has gone through some formal assessment of their quality system, such as SEMATECH's Standardized Supplier Quality Assessment (SSQA), (see Reference 1).
- Supplier has done its own self-assessment of the quality system

An alternative way to find out is to perform a formal supplier quality audit (assessment) of your own.

12.2 Communicate Reliability Requirements

The second step is to make sure that your supplier knows and understands the exact reliability requirements of the product you are going to purchase. These requirements at minimum should include the following:

1. Reliability level and metric (e.g., MTBF = 700 hours)
2. Time factor, such as the age of the equipment when it should attain the reliability level (e.g., 4 months after installation)
3. Operational conditions, such as:
   - Temperature and Humidity (e.g., temperature range: 70–75°F: Humidity range: 40–50% RH)
   - Duty cycle (e.g., 12 hours/day)
   - Throughput rate (e.g., 15 parts/hour)
   - Process to be used (e.g., high density plasma etch)
   - Operator skill level (e.g., grade 12 or equivalent)
   - PM policies to be followed (e.g., monthly PM policy described in the user's manual)
4. Shipping and installation limitations (e.g., to be shipped by air cushioned truck and installed by a special installation team)
5. Confidence level for the reliability metric (e.g., 80% confidence in the MTBF value)
6. Acceptable evidence for attaining the required reliability level (e.g., values based on in-house test data, or based on field data)

Every request for quotation (RFQ) and purchase order (PO) should include the above requirements. An even better idea is to develop a generic specification for the reliability requirements and include it with all the RFQ's and PO's.

12.3 Buy with Guarantee and Maintenance Contract

All RFQ's and PO's must include a guarantee clause that requires the supplier to guarantee the stated reliability level (e.g., MTBF = 500 hours). They should also include a penalty for not meeting the guaranteed reliability level. For example, if the guarantee level is not met, the supplier must provide free maintenance and spare parts.

If it is feasible, make the maintenance contract a part of the purchase agreement. This technique gives an added incentive to the suppliers to provide a reliable product. The suppliers profit from lower maintenance cost if they have high reliability.

12.4 Insist Upon a Reliable Product

Let your suppliers know that you are serious about the reliability requirements and insist that you will not take any less than the agreed upon reliability level. Before the product is shipped, make sure that your supplier provides credible evidences that the product meets its reliability requirements. If this is not feasible, perform a source inspection before the shipment.

12.5 Form Partnership with Suppliers

Establish a partnering relationship with your suppliers. Partnering is a business culture that fosters open communication and a mutually beneficial relationship in a supportive environment built on trust. To establish such a relationship, start at the top of both organizations. Begin frequent communications between technical groups. Employ a non-defensive approach to problem solving. Customer review the supplier's design for future generations of equipment or parts. Customer and suppliers participate in the formal design reviews and perform alpha or beta test on equipment.

Reference 2 is a good source of information for establishing a partnering relationship with your suppliers.

12.6 Provide Feedback

Provide detailed information to your supplier about all the problems and nonconformance you have observed during shipment, installation, start-up, and operations. Work with the supplier to eliminate these problems from future purchases. These problems may include problem analysis, failure analysis, and corrective action development and testing.

Complexity of this step depends upon the size of the supplier. For a small supplier, this step could be a simple information exchange. For a big supplier, this step could be a computerized FRACAS similar that described in Reference 3.

REFERENCES

1. SEMATECH Partnering For Total Quality, *Standardized Supplier Quality Assessment Workbook*, SEMATECH, Inc., Austin, TX, 1994.

2. Charles C. Poirier and William F. Houser, *Business Partnering for Continuous Improvements*, Quality Press, 1993.

3. SEMATECH, *Failure Reporting, Analysis and Corrective Action System*, Technology Transfer # 94042332A-GEN, SEMATECH, Inc., Austin, TX, 1994.

Chapter 13
Reliability Organization

So far, we have described the reliability improvement process and techniques. We have not mentioned anything about the organization and its people who deploy the improvement process and techniques. Even if the process and techniques are perfect, if they are not deployed properly by the proper person(s), the result may not amount to anything. It is important that we look at the organization and its people who drive the reliability discipline. This chapter describes the organization, its people, and their roles.

The reliability organization is a team of individuals who catalyze reliability improvement programs throughout an organization. They may not report to the same manager. The mission of this organization is to ensure that competitively priced equipment are manufactured which meet or exceed the customer's reliability requirements.

A reliability organization is not responsible for meeting the reliability goals of an equipment line (i.e., we cannot blame the reliability organization for failure to meet the equipment reliability goals). However, the organization has the responsibility for establishing and facilitating the right environment for the reliability improvement process. It is also responsible for providing technical knowledge and resources, taking initiative for reliability improvement tasks, and assigning ownership of the improvement process step.

13.1 Make-up of a Typical Reliability Organization

The typical reliability organization includes the following four levels of people:

1. Executive champion
2. Technical champion/reliability manager
3. Reliability engineer
4. Reliability technicians

Now let us describe who they are and what their duties are in more detail.

13.1.1 Executive Champion

The executive champion is a high-level person who could occupy any of the following upper management positions:

- President or vice president
- Chief operating officer
- Chief technical officer
- Corporate director of total quality

The role of the executive champion is to

- Provide executive leadership in reliability improvement matters
- Promote reliability improvement throughout the organization
- Work closely with the technical champion to develop the reliability improvement program plans
- Provide assurance that the reliability improvement programs are supported
- Mentor the reliability improvement programs and ensure that accomplishments are acknowledged

13.1.2 Technical Champion/Reliability Manager

The technical champion is a middle-level manager with thorough technical knowledge of reliability discipline, including reliability engineering and statistics. His or her duties are to

- Establish and facilitate the right environment for the reliability improvement process
- Provide both managerial and technical leadership
- Ensure deployment of effective cross-functional reliability improvement programs

- Ensure that the reliability improvement process is continuously applied throughout the life cycle phases
- Train other participants in reliability concept and improvement tools

13.1.3 Reliability Engineer

The reliability engineer is a degreed or experienced engineer with a thorough technical knowledge of reliability discipline, including reliability engineering and statistics. The duties of this engineer are as follows:

- Provide theoretical and practical tools to achieve the desired reliability level
- Provide technical leadership to specify, predict, design in, test, build in, manage growth of, and demonstrate the reliability level
- Manage the reliability growth until the desired reliability level is achieved
- Ensure deployment of effective cross-functional reliability improvement programs
- Ensure that the reliability improvement process is continuously applied throughout the life cycle phases
- Train other participants in reliability concept and improvement tools

13.1.4 Reliability Technician

The reliability technician is an experienced technician with thorough technical knowledge of operations and repair of the equipment and parts. The duties are as follows:

- Support reliability engineers
- Run reliability test
- Maintain the reliability test machines
- Collect reliability data from the field
- Perform failure analysis of the failed part(s)

13.2 Organization Structure

The location of the reliability organization within the total organization requires the following two major considerations: (i) provide management checks and balances designed to ensure that organization functions are kept in their true perspective, and (ii) ensure manufacturing of the equipment with a high-reliability level at an optimal cost.

Figure 13.1 depicts a typical organization chart that includes a reliability organization group. As shown in the chart, the reliability organization is under the director of engineering and reports to the director of operations with a dotted line structure. The organization position falls between the centralized and decentralized organizational structures. This structure appears to be very effective for the semiconductor manufacturing equipment suppliers.

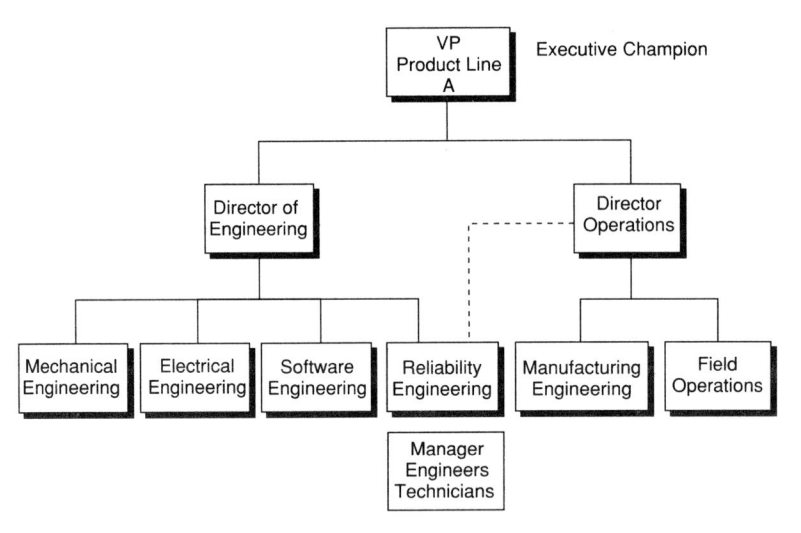

Figure 13.1 Typical Organization with a Reliability Engineering Group (Organization)

13.3 Recommended Practices for Reliability Engineers

Based on practical experience with many semiconductor manufacturing equipment suppliers, we recommend that reliability engineers

- Be flexible to react quickly to the demands of any reliability issue
- Train design, manufacturing, and field engineers in basic reliability methods and tools
- Be part of the design and design review teams

- Assist design engineers to design in reliability
- Assist manufacturing engineers to build in reliability
- Assist field engineers to collect the required data
- Coordinate failure review board activities

Chapter 14
How Good Is Your Reliability Improvement Program?

Today's highly competitive and global market environment requires an optimum level of reliability in every product that manufacturers make. To achieve this level, most manufacturing organizations have implemented some sort of reliability improvement program. However, they may not know how good their reliability improvement programs are. If you are a customer, you would definitely like to evaluate how well your key supplier's reliability improvement programs function. In this chapter, we will learn a simple methodology to evaluate the effectiveness of the reliability improvement programs.

Once the effectiveness is known, a manufacturer or a customer can use the results to:

- Satisfy curiosity
- Compare with a competitor or a benchmark company
- Identify weak areas in the program that may require some changes
- Gauge gradual improvement in effectiveness of the reliability improvement programs over time
- Market/purchase the equipment based on the effective reliability improvement programs in place

14.1 Evaluation Methodology

The evaluation methodology consists of answering a set of 58 questions related to the following seven categories:

1. Company culture, see Table 14.1
2. Reliability group (organization), see Table 14.2
3. Reliability goals/objectives, see Table 14.3

4. Design assurance, see Table 14.4
5. Reliability testing, see Table 14.5
6. Manufacturing quality assurance, see Table 14.6
7. Reliability growth management (continuous improvement), see Table 14.7

Table 14.1 Reliability Program Evaluation Checklist—Company Culture Related Activities

#	Company Culture Related Activities	Y/N
1.1	- Has the company conveyed the importance of reliability throughout the company?	
1.2	- Is the approach to reliability improvement proactive?	
1.3	- Is the product development process life-cycle oriented?	
1.4	- Does the company provide appropriate reliability training to staff involved with reliability improvement?	

1.5
1.6
	- Does the company have a partnering relationship with its
	    a. Customers?
	    b. Suppliers?	

1.7	- Does the company have a reliability data management methodology?	
1.8	- Does the company have or use a continuous improvement process to improve the reliability of its products?	
1.9	- Does the company use software quality and reliability improvement techniques to improve software reliability?	
1.10	- Does the company use emerging new technology (hardware and software) in its equipment designs to improve reliability?	

Table 14.2 Reliability Program Evaluation Checklist—Reliability Group Related

#	Reliability Group Related	Y/N
2.1	- Does the company have a formally organized reliability group?	
2.2	- Are the group members' responsibilities defined?	
2.3	- Is there a formal reliability program plan?	
2.4	- Has the plan been approved and resources committed by the high level managers?	
2.5	- Has the reliability plan produced any results?	

Once the answers are collected and tallied, Figure 14.1 is used to evaluate the standing of an organization with respect to an average organization and to a benchmark organization.

14.2 Evaluation Process Steps

1. Select a group of people who have knowledge of the organization for which the reliability improvement programs are being evaluated.
2. Have the selected individuals fill out the questionnaire form. Their answers should be either 'Yes' or 'No' for each question.
3. Tally the results and calculate the average numbers of 'Yes' occurrences for each question, each category, and overall.
4. Use your overall score and Figure 14.1 to find out where the organization stands with respect to an average organization and benchmark organizations. This scale is very generic. You can develop your own scale based on your surveys for many organizations.

For example:

If your average overall score is 28, your reliability improvement program is average—similar to that of an average organization.

If your average overall score is 52, your reliability improvement program is a benchmark program.

Table 14.3 *Reliability Program Evaluation Checklist—Reliability Goals/Objective Related Activities*

#	Reliabilty Goals/Objectives Related Activities	Y/N
3.1	- Does the company understand its customers' reliability requirements?	
3.2	- Does the company establish product-specific reliability goals/objectives?	
3.3	- Does the company use sound methods (Cost of Ownership, benchmarking, customers' requirements, etc.) to establish reliability goals/objectives?	
3.4	- Are reliabilty goals adequately defined? (At minimum they include a reliability metric, confidence level, intended use conditions, and age of the equipment.)	
3.5	- Are the reliability goals realistic and achievable?	
3.6	- Does the company involve customers in establishing reliability goals?	

3.7
3.8
3.9
	- Does the company use a formal process to allocate the system level reliability goals to :
	    a. Hardware and software?
	    b. Subsystem level?
	    c. Component level?	

3.10	- Does the company understand its customers' environment, health and safety requirements?	

Table 14.4 Reliability Program Evaluation Checklist—Design Assurance Activities

#	Design Assurance Activities	Y/N
4.1	- Is there a formally organized review team that addresses the reliability requirements?	
4.2	- Are design reviews held on a regular basis?	
4.3	- Is there a formal engineering change control process in use and is it effective?	
4.4	- Does the company use designing-in reliability techniques such as design simplifications, part count reduction, derating and redundancy?	
4.5	- Does the company use "Design For Manufacturabilty" techniques to optimize the production cost and reliability?	
4.6	- Does the company use more standard (proven) parts than new parts?	
4.7	- Does the company specify part reliability requirements on drawings?	
4.8	- Does the company perform reliability modeling to assess design?	
4.9	- Does the company perform Failure Mode and Effects Analysis (FMEA) and/or Fault Tolerance Analysis (FTA)?	
4.10	- Does the company perform subsystem and part-level life tests?	
4.11	- Is there a formal procedure in place to transfer design to the prototype phase?	

Table 14.5 Reliability Program Checklist—Reliability Testing Tasks

#	Reliability Testing Tasks	Y/N
5.1	- Does the company prepare the test plans with -- statistically adequate sample size and test length? -- operating conditions? -- pass/fail criteria? -- definition of a failure? -- data needs? -- data analysis procedure? -- failure reporting and correcting procedure? -- test facility?	
5.2	- Does the company conduct the test according to the test plan?	
5.3	- Are the test results analyzed and conclusions made?	
5.4	- Are the failures analyzed and reported to the proper parties for corrective actions?	

Table 14.6 Reliability Program Evaluation Checklist—Manufacturing Quality Assurance Tasks

#	Manufacturing Quality Assurance Tasks	Y/N
6.1	- Are the supplier selection and approval procedures (based on quality and reliability of their products) in place?	
6.2	- Does the company adequately communicate reliability requirements to the part suppliers? (At minimum these include a reliability metric, confidence level, intended use conditions, and age of the parts.)	
6.3	- Is there a formal procedure in place for controlling materials and manufacturing processes?	
6.4	- Are the assembly inspection procedures defined and performed?	
6.5	- Is there a formal procedure to feed back the manufacturing problems to design engineering?	
6.6	- Does the company conduct Product Reliability Acceptance Tests (PRAT)?	
6.7	- Are the shipping and handling procedures defined and performed?	
6.8	- Does the company provide training in manufacturing procedures to the manufacturing labor force?	

Table 14.7 Reliability Program Evaluation Checklist—Reliability Growth Management Tasks

#	Reliability Growth Management Tasks	Y/N
7.1	- Does the company maintain a product configuration matrix and the customer configuration logs?	
7.2	- Is there a system for deploying and communicating reliability improvement changes?	
7.3	- Does the company identify and correct problems observed during shipping and installation?	
7.4	- Does the company's training for operation and maintenance of equipment include reliability improvement tasks such as preventive maintenance?	
7.5	- Does the company provide adequate technical documents for operation and maintenance of the equipment and software?	
7.6	- Is there a formal field Failure Reporting, Analysis and Corrective Action System (FRACAS) in operation?	
7.7	- If the reliability goals are not met, is there a formal reliability growth plan in place?	
7.8	- Is there a formal procedure in place to transfer the field experience (such as the component failures, inherent problems, etc.) to the future generations of equipment?	
7.9	- Is the component failure rate data communicated to respective suppliers?	
7.10	- Are the suppliers held accountable for not meeting the respective reliability goals?	

| | RANGE OF OVERALL 'YES' SCORED |||||
	0-20	21-30	31-40	41-50	51-58
STANDING OF RELIABILITY PROGRAM	Below Average	Average	Above Average	Excellent	Benchmark Level

Figure 14.1 Reliability Program Evaluation Guide

Appendix

Table A1. Concept and Feasibility Phase: Activities by Reliability Improvement Process Step

Reliability Improvement Process Step	Activities
1. Establish Goals and Requirements	• Establish reliability goals • Create reliability program plan
2. Reliability Engineering and Improvements	• Develop functional block diagrams & preliminary reliability model • Allocate reliability goals • Collect historical failure data • Develop preliminary Life Cycle Cost
3. Conduct Evaluation	• Preliminary prediction of equipment reliability • Conceptual design review(s)

4. Are Goals and Requirements Met?	• Compare goals to predicted reliability values
	  - if goals are not met, continue to Step 5
	  - if goals are met move to design phase of life cycle

5. Identify Problems and Root Causes	• Perform sensitivity analyses using reliability model

Table A2. Design Phase: Activities by Reliability Improvement Process Step

Reliability Improvement Process Step	Activities
1. Establish Goals and Requirements	• Modify goals to match customer requirements • Update reliability program plan
2. Reliability Engineering and Improvements	• Apply design-in-reliability practices • Expand reliability model to include more detailed subsystems • Allocate subsystem requirements to subsystem components • Collect failure data for components within subsystems • Evaluate reliability of purchased components • Run life test on new and critical components • Update Life Cycle Cost • Perform ergonomics and human factors studies • Conduct software reliability studies • Implement failure mode and effects analysis (FMEA)
3. Conduct Evaluation	• Assess equipment reliability using reliability modeling • Conduct design reviews

4. Are Goals and Requirements Met?	• Compare reliability requirements to predicted values
	  - If requirements are not met, continue to Step 5
	  - If requirements are met, move to prototype phase of life cycle

5. Identify Problems and Root Causes	• Perform sensitivity analyses to identify problems

Table A3. Prototype Phase: Activities by Reliability Improvement Process Step

Reliability Improvement Process Step	Activities
1. Establish Goals and Requirements	• Update reliability requirements • Update reliability program plan
2. Reliability Engineering and Improvements	• Update reliability model, as needed • Re-allocate subsystem and component reliability requirements • Establish test plan • Establish FRACAS • Perform human reliability analysis • Develop preventive maintenance program • Perform ergonomics studies • Conduct software reliability studies • Update Life Cycle Cost
3. Conduct Evaluation	• Conduct tests of prototype equipment and evaluate reliability • Conduct design review(s)

4. Are Goals and Requirements Met?	• Compare reliability requirements to predicted values
	  -If requirements are not met, continue to Step 5
	  -If requirements are met, move pilot production phase of life cycle

5. Identify Problems and Root Causes	• Perform sensitivity analysis • Evaluate FRACAS to identify problems and root causes • Evaluate FMEA to identify potential failure modes • Perform failure analysis on critical components

Table A4. Pilot Production Phase: Activities by Process Improvement Step

Reliability Improvement Process Step	Activities
1. Establish Goals and Requirements	• Update reliability requirements as needed • Update reliability program plan
2. Reliability Engineering and Improvements	• Upgrade testing program as needed • Implement FRACAS, if not already done • Perform human reliability analyses • Perform software reliabiliy studies • Perform ergonomic studies • Update preventative maintenance program, as needed • Update Life Cycle Cost
3. Conduct Evaluation	• Conduct tests and evaluate reliability • Conduct design review(s)

4. Are Goals and Requirements Met?	• Compare reliability requirements to observed values
	  - If requirements are not met, continue to Step 5
	  - If requirements are met move to production & operations phase of life cycle

5. Identify Problems and Root Causes	• Perform sensitivity analyses • Evaluate FRACAS • Perform failure analyses on critical components

Table A5. Production and Operation Phase: Activities by Reliability Improvement Step

Reliability Improvement Process Step	Activities
1. Establish Goals and Requirements	• Final update of reliability requirements, if needed • Final update of reliability program plan
2. Reliability Engineering and Improvements	• Use FRACAS for field tracking, customer feedback and corrective action program • Update human reliability analyses • Update software reliabiliy studies • Update ergonomic studies • Update preventative maintenance program, as needed • Continue to evaluate reliability of purchased components • Update Life Cycle Cost, if required
3. Conduct Evaluation	• Assess equipment reliability based on the field data • Evaluate feedback from field tracking and maintenance records

4. Are Goals and Requirements Met?	• Compare reliability requirements to observed values
	  - If requirements are not met, continue to Step 5
	  - If requirements are met:
	    * Continually monitor equipment performance
	    * Implement process of continuous improvement
	    * Eventually phase out current generation equipment

5. Identify Problems and Root Causes	• Perform failure analyses on field failures

Table A6. Phase-Out Phase: Activities by Reliability Improvement Step

Reliability Improvement Process Step	Activities
1. Establish Goals and Requirements	• Set requirements for subsystems and components to be carried to the next generation of equipment • Document and retain all information gathered during generation of equipment being phased out
2. Reliability Engineering and Improvements	• Offer phase-out alternatives to customers of equipment being phased out • Phase out current generation equipment in stages
3. Conduct Evaluation	• Assess reliability of the current generation and carried information to next generation of equipment
4. Are Goals and Requirements Met?	• There are no goals or requirements to meet
5. Identify Problems and Root Causes	• Retain all information on equipment being phased out so that it can be used in future generations of equipment

Index

A

Accelerated Test 99
acceptable evidence 76
acquisition cost 52
AGREE 1, 77
AITPM 50
allocation 76
   equal 77
   goal 76
   method, AGREE 77
   method, ARINC 77
alpha-site evaluation 63
application
   intended 79
   of reliability metrics 27
apportionment 76
ARINC 77
ASQC 1
assembly instructions 87
assessed
   reliability 28
   value 30
assessment
   supplier 3
availability 26, 46

B

bathtub curve 18
beta-site evaluation 63
better equipment reliability 75
bite-size goal 76

budgeting 76
build-in 59
build-in reliability 87
burned-in 87
burning 19
buying 105
buying process 105

C

calculations
   reliability 19
CDF 5, 13, 14
champions
   reliability 72
chance failure 19
chief operating officer 112
chief technical officer 112
closed-loop 87
company culture 117
company-level plan 71
component
   nonrepairable 5
   repairable 5
concept
   and feasibility phase 60
conditions
   operational 76
   stated operational 6
confidence level 76, 103
CoO 3, 49

corporate director 112
corrective actions 69, 88
corrosion 19
cost
  acquisition 52
  fixed 55
  life cycle 2, 51
  of reliability 2
  operating 55
  operational 52
  waste disposal 54
  yield loss 55
Cost of Ownership 49, 54
creep 19
critical failure 19
CT
  actual 50
  theoretical 50
customer service 62
cycle time 51

D

data analysis techniques 100
debug test 98
definition
  for reliability 5
dependent failure 19
derating 62
design
  engineer 2, 62
  phase 60
  review 62
  simplification 81
  software 82
design techniques 78
designed-in 2
design-in 59
designing-in 62

desired
  confidence level 103
  value 27, 28, 29
duty cycle 76

E

early
  failure 18, 19
  life 60
electromagnetic compatibility 82
empowered 90
engineer
  design 2, 62
  field service 2
  manufacturing 2, 62
  reliability 62, 112
engineering time 56
Environmental Stress Screening Test 98
equal allocation 77
equipment 105
  life cycle phase 60
  military 1
  performance 33
  program 59
  states 33
evaluation
  alpha-site 63
  beta-site 63
  methodology 117
  process 120
executive champion 112
expected outcome 100
expected/predicted reliability 28
expenses 54
explosion 82
exponential PDF 23
external factors 78, 81

F

failure
　catastrophic 18
　categories 5, 17
　chance 18, 19
　critical 19
　data reporting 88
　degradation 18
　dependent 19
　early 18, 19
　independent 19
　intermittent 18
　nonrelevant 19, 20
　rate 23
　rate function 12
　relevant 19
　time 20
　wear-out 18, 19
Failure Review Board 88
fatigue 19
feasibility
　and concept phase 60
feedback 63, 109
field service
　engineer 2
fixed cost 55
FRACAS 87
FRU 21
function
　failure rate 5, 12
　intended 6, 17, 19, 20
　probability 13
　reliability 5, 7, 11, 12, 13

G

Generic Steps for Reliability Test 100
goal 33, 75
　allocation 76
　bite-size 76
　subsystem level 28
　value 32
Growth Test 99
guarantee 108
Guideline E10-92 26

H

heat generation 82
high
　temperature 82
　vacuum 82
high-level equipment
　performance metrics 49
human use 82

I

IEEE 1
independent
　failure 19
inferred 17
inherent reliability 28
inspecting 2
installation 76
insurance 54
intended
　function 19, 20
ISO 9001 107

K

K Factor 104

L

LCC 49
level
　confidence 76, 103
life
　cycle cost 2, 51
　cycle phase 59

Index　　135

cycle, equipment phase 60
early 60
measure of 7
operational longevity 45
test 8
useful 60
wear-out 60
Life Cycle Cost 49
life test 9
longevity
  operational life 45
long-term strategy 72
lower confidence limit 103

M

maintainability 45
  and reliability 45
maintenance
  periodic preventive 82
  predictive 83
  procedure 6
  scheduled 50, 82
  unscheduled 50
Malcolm Baldrige National Quality Award 107
manager
  reliability 112
managers
  marketing 2
  program 2
  purchasing 2
managing reliability growth 59, 69
manufacturing
  engineer 2, 62
marketing 62
Mean Cycles Between Down Events 26
mean life 24, 25
Mean Time Between Failures 26
Mean Time To Repair 45
Mean Wafers Between Failures 26

measure of life 7
mechanical stops 81
mechanisms
  reliability growth 91
methodology
  evaluation 117
metrics
  application of reliability 27
  applications 23
  hierarchy of equipment performance 57
  high-level equipment performance 49
  precise use of reliability 32
  reliability 23, 33
  standardization of reliability 33
modeling
  predictive 69
modules 76
moisture 82
MTBF 23, 32, 33
  observed 103
MTTR 35, 45

N

nonrelevant failure 19, 20
non-repairable system 20

O

observed
  value 27, 29, 30, 32
observed MTBF 103
OEE 3, 45, 49
operating
  cost 55
  environment 6
  speed 6
  stress level 6
operational
  conditions 76
  cost 52

operator
　skill level 6, 76
other disciplines 43
Overall Equipment Effectiveness 49

P

packaging
　and shipping 87
parallel system 37
part
　derating 79
　electronic 1
　selection 79
　specification 79
partnering 2
partnership 108
pass-fail criteria 100
PDF 5, 13, 14
　exponential 14
percent failed 23
percentages 24, 27
performance
　efficiency 50
periodic preventive maintenance 82
phase
　concept and feasibility 60
　design 2, 60
　development 2
　equipment life cycle 60
　life cycle 59
　phase out 60
　pilot production 60
　product life cycle 1
　production 60
　prototype 60
phase out phase 60
pilot production phase 60
plan
　company-level 71
　improvement 3

　product-level 71
　two-tier reliability 71
PO 108
policy
　reliability 72
population 5, 16, 17
PPH
　actual 50
　theoretical 50
PRAT 87
predictive
　maintenance 83
　modeling 69
president 112
pressure regulators 81
probabilistic metrics 25
probabilities 24
probability
　density 14
　function 13
　of success 25
　of survival 25
procedure
　maintenance 6
　screening 79
process
　buying 105
　evaluation 120
　independent renewal 41
　reliability improvement 67
Product Reliability Acceptance Test 87
production
　phase 60
　volume 54
product-level plan 71
program
　equipment 59
　improvement 2
　joint development 3
　management 1

protective technique 81
prototype phase 60

Q

Qualification Test 99
quality
   and reliability 43
   assurance personnel 63
   control 43
   rate of 50, 51
   system 107

R

random variable 5
rate
   of quality 50, 51
   throughput 55
redundancy 62, 81
relevant failure 19
reliability
   and maintainability 45
   assessed 28
   basic elements 5
   better equipment 75
   build-in 87
   calculations 19
   champions 72
   cost of achieving 2
   cross-functional 112
   definition 5
   discipline 1
   engineer 62, 112
   engineering 1, 2
   expected/predicted 28
   formal courses 2
   formal program 1
   function 7, 11, 12, 13
   goal 24
   growth mechanisms 91

   improvement process 67
   improvements 2
   inherent 28
   management of growth 59
   manager 112
   managing growth 69
   metrics 23, 33
   of systems 37
   optimal level of 52
   optimum 2
   policy 72
   precise use of metrics 32
   proper level of 49
   quality and 43
   safety and 45
   skills 72
   software 2
   standardization of metrics 33
   system-level 38
   technicians 112
   testing 97
   textbooks 1
   two-tier plans 71
   types of tests 98
reliability organization 111
repairable system 20
reporting
   failure data 88
request for quotation 108
requirements 3, 75
RFQ 108
root causes 67

S

safety 45
   and reliability 45
sample size 100
samples 5, 16, 17
scheduled maintenance 82
scrap loss 54

screening 2, 19
self-assessment 107
SEMATECH 55
SEMI 26
SEMI E10-92 33, 35
sequential 60
series 21
shipping 76
   and packaging 87
shock and vibration 82
skill level
   operator 6
skills
   reliability 72
societies
   professional 2
software 20
   design 82
SSQA 107
standardization
   of reliability metrics 33
standby 21
   system 40
   time 56
strategy
   long-term 72
supplier 106
   certification 3
supplier assessment 3
system
   non-repairable 20
   nonrepairable 5
   pure parallel 40
   quality 107
   repairable 5, 20, 37, 41
   series 37
   standby 37, 40
systems
   reliability of 37

T

taxes 54
technical champion 112
temperature
   high 82
test
   Accelerated 99
   conducting 100
   data analysis 103
   data analysis and reporting 100
   debug 98
   Environmental Stress Screening 98
   Generic Steps for Reliability 100
   Growth 99
   length 100, 102
   life 8, 9
   plan development 100
   procedure 100
   Qualification 99
   reliability, types of 98
theoretical CT 50
theoretical PPH 50
thermostats 81
throughput 76
   rate 55
   yield 55
time 7
   engineering 56
   factor 76
   failure 20
   productive 33
   specified 6
   standby 56
   total calendar 50
   unscheduled maintenance 50
Total Productive Maintenance 49
total quality 112
TPM 49
training 87

U

U.S. Department of Army 1
U.S. Department of Navy 1
units
   field replaceable 21
uptime 46
useful life 60

V

vacuum
   high 82
value
   analytical/theoretical 27, 28
   assessed 30
   desired 27, 28, 29
   goal 32
   observed 27, 29, 30, 32
   present 54
variable
   random 5
vice president 112
volume
   production 54

W

waste disposal cost 54
wear-out
   failure 18, 19
   life 60
Weibull distribution 92
worth 54

Y

yield
   loss cost 55
   throughput 55